STUDIES IN APPLIED MATHEMATICS

Studies in Mathematics

The Mathematical Association of America

G. F. Carrier
Harvard University

Hirsh Cohen
IBM Watson Research Center, Yorktown Heights

Donald Greenspan
University of Wisconsin, Madison

Tosio Kato
University of California, Berkeley

C. C. Lin
Massachusetts Institute of Technology

A. H. Taub
University of California, Berkeley

G. B. Whitham
California Institute of Technology

Studies in Mathematics

Volume 7

STUDIES IN APPLIED MATHEMATICS

A. H. Taub, editor
University of California, Berkeley

Published by
The Mathematical Association of America

Distributed by
Prentice-Hall, Inc.

Current printing (last digit):

10   9   8   7   6   5   4   3   2   1

PREFACE

Applied mathematics is concerned with the understanding of scientific phenomena by means of the construction, analysis, and interpretation of mathematical models. Particular problems may arise not only in the areas of the classical physical and engineering sciences but also in the social, computing, biological, medical, and management sciences. Construction and interpretation especially depend not only on mathematical ideas, abstractions, and methods but also on an understanding of the principles, methods, and practice of the area in which the problem arises. However, an applied mathematician may be said to differ from a theoretician working in a given scientific field in that the former is more likely to treat a given problem as a case history of a particular method of mathematical analysis than is the latter.

The applied mathematician's success depends not only on his mathematical ability but equally on his ability to formulate idealized but relevant mathematical models of a class of phenomena and to pose and interpret the answers to precise and cogent mathematical questions of the model which on one hand have a likelihood of being answered and on the other hand are relevant to the understanding of the situation being analyzed. The reduction of a complex problem to a model whose analysis involves elementary mathematics, and which leads to results that may be verified by means of the data available is exciting applied mathe-

matics in spite of the fact that the mathematics used in not novel.
This situation does not occur often. It is more often the case that
the analysis of the models brings new mathematical problems that
require the development of new mathematical techniques for their
solution. New branches of mathematics have been created in the
past to deal with problems that originated in areas outside of
mathematics.

This volume of the MAA studies is intended to bring to the
attention of advanced undergraduate students, graduate students,
and other members of the Association some current work in applied
mathematics. The papers are drawn from a relatively small segment of this subject. All of them deal with problems which either
originate in the physical sciences or for which a model has been
created based on a problem in such a science. It should be emphasized that the papers are not representative of all or the major
portion of the work now being done in applied mathematics. It
is to be hoped that subsequent volumes of the MAA Studies will
bring forth expository articles on other parts of applied mathematics.

The article by G. F. Carrier is a reprint of his John von Neumann
lecture delivered at the 1969 SIAM Summer Meeting, August
1969 at Eugene, Oregon. In this paper the three steps—model
creation, mathematical analysis, and interpretation—are well illustrated by his description of attempts to understand some perplexing phenomena in the dynamics of oceans and atmospheres
by the use of models whose analysis requires the use of singular
perturbation theory. The paper begins with a brief description
of that theory and then applies the theory to two examples: Mid-
latitude ocean circulation and hurricanes.

H. Cohen's paper provides an example of the "case history"
aspect of applied mathematics. In this paper he discusses the
manner in which nonlinear diffusion equations differ from linear
ones and provide quite novel phenomena. In particular, the smear-
ing out of amplitudes and other features is shown not to occur;
instead, traveling wave or other frontal motions appear. The examples where such problems arise are drawn from physiology,
chemistry, physics, and geophysics.

In the third paper D. Greenspan explores some of the effects that computers are having on applied mathematics by discussing the discrete approach to physical problems, the approximate solution of unsolved continuous linear problems, and the approximate solution of unsolved continuous nonlinear problems. The discussion of the examples used to illustrate problems of continuous linear and nonlinear types shows how the power of a modern computer can be used to obtain approximate solutions of problems whose complexity is such that present-day mathematics cannot provide the information desired concerning these problems. The discussion of the discrete approach to physical problems illustrates another aspect of the application of computers that is potentially more important than their use as glorified slide rules. This aspect is the following one: New models of physical (and other) phenomena may be created that are more adapted to computer analysis than are those of current physical theories. Thus some scientists who are aware of the nature and properties of computers prefer to consider problems in the kinetic theory of gases (and continuum mechanics) in terms of a model made up of a collection of particles that interact with each other by collision or other mechanisms. The computer is given the task of treating these particles individually instead of dealing with approximations to continuous functions obtained by averaging over the particles. Thus the computer treats a discrete model close to the actual model instead of a discrete approximation to a continuous approximation of a discrete model. It seems likely that computers will play an increasingly important role in science by providing a means for dealing with novel types of models—types which are not suitable for the application of present-day mathematical techniques but which are suitable for treatment by computers.

T. Kato's paper discusses a concrete and simple example in scattering theory and thereby illustrates the large body of mathematics into which this theory has grown. The problem discusses the nonrelativistic quantum mechanical description of the scattering of an electron by a potential field V. The article demonstrates the manner in which the theory of linear operators in Hilbert space enters into the discussion of quantum mechanical problems and how such problems have influenced the mathematical development of that theory.

The article by C. C. Lin is also a reprint of a John von Neumann SIAM lecture, the eighth one delivered at Toronto, Ontario on August 30, 1967. In it the mathematical theory of a system of stars in the form of globular clusters or of galaxies is treated. The former system consists of hundreds of thousands of stars and the latter one of hundreds of billions of stars. A statistical theory pattern after that of atomic stochastic processes is described, but here the stars play the role of atoms. The discussion treats the behavior of the stars both individually and collectively through the use of ideas from the Maxwell-Boltzmann approach to kinetic theory. Thus the "case-history" aspect of applied mathematics is again illustrated in still another area. The author also emphasizes the interplay between observation and theory. This treatment of the nature of collective modes and relaxation processes in a "collisionless" system stimulates further observations and points to the need for the development of a general mathematical theory of nonlinear random processes.

The writer's paper deals with the extension of classical hydrodynamics to the special and general theories of relativity. The classical theory is of interest not only because it is related to many problems that arise in the physical and engineering sciences but also because its mathematical structure involves nonlinear hyperbolic partial differential equations. The mathematical theory of such equations is in a very rudimentary stage compared to the theory of linear elliptic partial differential equations. The fact that the equations model the behavior of a physical system enables one to acquire, from the knowledge of the behavior of such systems, some intuition regarding the mathematical properties of the solutions of nonlinear hyperbolic partial differential equations.

It is of some theoretical interest to determine whether it is possible to modify the classical theory to bring it in accord with the underlying postulates of the theory of relativity, that is, to formulate it in such a way that it is invariant under the inhomogeneous Lorentz group instead of the galilean group. There are some practical consequences that emerge from these considerations. The generalization to general relativity has received considerable interest of late because of the application being made of the theory to discussions of astronomical problems such as the gravitational collapse of massive bodies and the large-scale motions that can take place in the universe. General relativistic hydrodynamics treats the theory of self-gravitating fluids and is a combination of two nonlinear theories, each of which is poorly understood from a mathematical point of view, for each one involves complicated nonlinear partial differential equations.

The final paper by G. B. Whitham reviews the recent developments in the theory of linear and nonlinear dispersive wave motion. Such motions are exemplified by motions in a medium when local disturbances disperse into a wave train. Problems from a variety of disciplines are presented from a unified point of view, which is based on the use of variational principles. This paper gives another example of the "case-history" approach to the applied mathematician. It also illustrates the power of a unifying principle. Problems from a variety of fields are brought together and analyzed by one method that applies to both linear and nonlinear problems. In the discussion, the notions of phase velocity of waves and the group velocity are clearly described and their relationship is examined. The use of variational principles enables one to discuss stability problems and the behavior of perturbations. The methods of analysis used in this paper have a great range of application and will be useful in the treatment of very many problems arising in diverse fields.

A. H. Taub

ACKNOWLEDGMENTS

Professor Carrier's article, "Singular Perturbation Theory and Geophysics," has been reprinted with permission from the SIAM REVIEW, Vol. 12, No. 2, 1970. In Dr. Cohen's article, "Nonlinear Diffusion Problems," figures 5, 6, and 7 have been reprinted with permission from the JOURNAL OF CELLULAR AND COMPARATIVE PHYSIOLOGY, Vol. 66 (Supplement 2), 1965; figures 12 and 13 have been reprinted with permission from the JOURNAL OF MATHEMATICAL PHYSICS, Vol. 3, 1962; figures 16 and 17 have been reprinted with permission from the QUARTERLY JOURNAL OF MECHANICS AND APPLIED MATHEMATICS, Vol. XVII, No. 2, pp. 141–155, 1964. Professor Lin's article, "Dynamics of Self-Gravitating Systems: Structure of Galaxies," has been reprinted with permission from the SIAM REVIEW, Vol. 11, No. 2, 1969.

CONTENTS

STUDIES IN APPLIED MATHEMATICS

SINGULAR PERTURBATION THEORY AND GEOPHYSICS

G. F. Carrier

1. INTRODUCTION

Were it not for the usual compromise with brevity, the title of this lecture might have been "A brief review of singular perturbation theory as mathematics, some perplexing phenomena in the dynamics of oceans and atmospheres, and the role of singular perturbation ideas in attempts to understand these phenomena."

The selection of questions from geophysics to illustrate areas of application of singular perturbation theory is not merely a result of the fact that such techniques happen to be useful in that science. The understanding of geophysical fluid mechanics is at a stage where the appropriate mathematical models have not been clearly delineated, and partial successes in the evolution of such models and in the understanding of the obstacles encountered are tied closely to the ideas underlying the techniques. Accordingly, geophysics seems to be a better vehicle for an exposition of the interplay between mathematical ideas and scientific understanding than are other areas of application in which the mathematical

1

formulations already are well authenticated and in which the mathematical substance is used *only* to obtain quantitative descriptions.

It is easier to elaborate on this composite topic through illustrative examples than it is to give a comprehensive account of either the mathematical state of the art or of the state of understanding of geophysical fluid mechanics. Accordingly, after a characterization of what is meant by "singular perturbation problems," we shall contrast (for each of two pairs of problems) a situation whose understanding is backed by rigorous analysis and a very similar situation whose understanding is based on heuristic arguments. This should give a fairly clear picture of the boundary between mathematical questions that are still open and those that have been answered. The details of proof and technique will be alluded to only by reference to the literature.

In order to characterize the extent to which geophysical dynamics problems are understood, two examples will be used again. The particular phenomena are chosen not only because they draw heavily on the use of singular perturbation methods but also because they illustrate, particularly clearly, the nature of the difficulties that impede progress in geophysical understanding and the nature of the exploratory mathematical models that are so useful in the study of individual facets of the phenomena.

2. SINGULAR PERTURBATION THEORY

When one encounters a boundary value problem (or an integral equation or an algebraic problem) that contains a parameter α and when the behavior of its solution is of interest for very small value of α, one is tempted to seek a solution in the form of an ascending series in α. When such attempts fail, the problem is said to be a singular perturbation problem; singular perturbation *theory* deals with questions concerning the descriptions of solutions of such problems.

The nature of the failure can vary. For example, in the problem†

---

† Throughout this paper a subscript S denotes the partial derivative with regard to the variable S.

(2.1) $$\epsilon u_{xx}(x, \epsilon) - (2 - x^2)u(x, \epsilon) = -1$$

in

$$-1 < x < 1, \qquad 0 < \epsilon \ll 1,$$

with

$$u(-1, \epsilon) = u(1, \epsilon) = 0,$$

the solution has an asymptotic description

(2.2) $$u = \sum u_n(x)\epsilon^n$$

in a region that does *not* include the end points, but no description of this form describes the solution in $1 \geq |x| > 1 - O(\epsilon^{1/2})$. Thus, there is no a priori assurance that the formal construction of such a description will lead to a correct result, because its range of validity cannot possibly contain those points at which the solution must be consistent with the prescribed boundary value information.

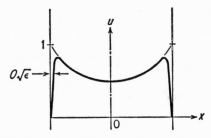

FIG. 1. Solution of problem associated with Eq. (2.1). Note the role of ϵ in the structure of the solution.

A different sort of failure occurs when one attempts a description of the form of Eq. (2.2) for the problem

(2.3) $$u_{xx}(x, \epsilon) + \epsilon[u_x(x, \epsilon)]^3 + u(x, \epsilon) = 0$$

in

$$0 < x < \infty, \qquad 0 < \epsilon \ll 1,$$

with

$$u(0, \epsilon) = 1, \qquad u_x(0, \epsilon) = 0.$$

In this example, the series is easily constructed, and it converges in a region $0 \leq x < R = 4/3\epsilon$. However, it fails either to converge to $u(x, \epsilon)$ or to describe it asymptotically, in $x > R$. *More impor-tant,* unless one manages to sum the series so constructed, it fails to be interpretable when $x \geq O(1/\epsilon)$. Since this is the very region in which the behavior of u is of interest, the expansion fails to serve the desired purpose.

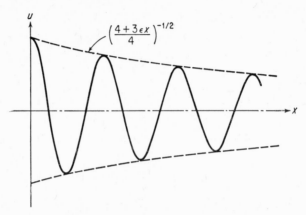

FIG. 2. Solution of problem associated with Eq. (2.3). Note the role of ϵ in the structure of the solution.

Finally, in the problem

$$(2.4) \quad u_{xx}(x, \epsilon) + \epsilon\beta u_x(x, \epsilon) + [\tfrac{1}{4} + \sigma\epsilon + \epsilon \cos x]u(x, \epsilon) = 0$$

in

$$0 \leq x < \infty, \quad 0 < \epsilon \ll 1,$$

with

$$u(0, \epsilon) = 1, \quad u_x(0, \epsilon) = 0,$$

$u(x, \epsilon)$ is an entire function of ϵ, but once again its power series description (in powers of ϵ) is entirely uninformative in the context of the scientific problems from which it typically arises.

The solutions of the three problems just cited are superbly approximated, respectively, by

$$(2.5) \quad u^{(1)} \sim \frac{1}{2 - x^2} - e^{-(x+1)/\sqrt{\epsilon}} - e^{-(1-x)/\sqrt{\epsilon}},$$

$$(2.6) \quad u^{(2)} \sim (1 + 3\epsilon x/4)^{-1/2} \cos x,$$

and

$$(2.7) \quad u^{(3)} \sim e^{-\epsilon\beta x/2} \left[\cosh \nu x \epsilon \cos x/2 + \sqrt{\frac{1-2\sigma}{1+2\sigma}} \sinh \nu x \epsilon \sin \frac{x}{2} \right],$$

where $\nu = (1/4 - \sigma^2)^{1/2}$. Graphs of each are shown schematically in Figs. 1 to 3.

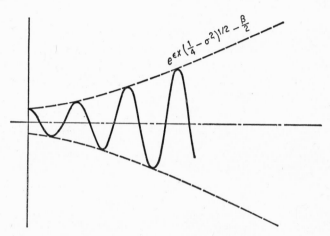

FIG. 3. Solution of problem associated with Eq. (2.4). Note the role of ϵ in the structure of the solution.

This elaborate presentation of the solutions is included for two reasons:

1. It illustrates those features which explain the "failures" noted in the foregoing. In particular, with reference to Eq. (2.5), $e^{-(x+1)/\sqrt{\epsilon}}$ is irrelevant to an asymptotic description in $x \geq a > -1$, but it is essential to any description of what happens in $x = -1 + O(\sqrt{\epsilon})$; in this region, of course, it does not permit a description having the form of Eq. (2.2). The importance of this term to the description of the general features of $u^{(1)}$ is amply illustrated in Fig. 1.

The statements about the convergence and interpretability of series expansions of $u^{(2)}$ and $u^{(3)}$ are easily seen to be consistent with Eqs. (2.6) and (2.7).

2. It is clearly evident both from Eqs. (2.5), (2.6), and (2.7) and from Figs. 1, 2, and 3 that different facets of each solution are described in terms of different variables. For example, $u^{(1)}$ is characterized in much of the region by a "smooth" behavior, which is a function of x, but appended to this are *steep* boundary layers, which intrinsically are functions of $(x - x_{\text{boundary}})/\epsilon^{1/2}$. Alternatively, $u^{(2)}$ and $u^{(3)}$ display their oscillatory behavior through a function of x, but their decay (or growth) is characterized by a function of ϵx. This *multiscale* character of $u^{(1)}$, $u^{(2)}$, and $u^{(3)}$ is a characteristic feature of the functions that describe the solutions to singular perturbation problems.

Rather formal presentations of the various techniques by which such problems can be treated and many examples thereof are given in [1], [2], and [5] through [10]. Some informal presentations are given in [3] and [11] through [13].

Under appropriate definitions of "dominance," a rigorous justification of the statement "$u^{(1)}$ and $u^{(3)}$ are dominated uniformly by the descriptions given in Eqs. (2.5) and (2.7)" is within the state of the art. I know of no *proof* of the corresponding statement for Eq. (2.6).

In summary, then, singular perturbation problems are characterized by the fact that their solutions exhibit a multiscale "behavior" that depends on the crucial role played by a small parameter in the defining equations of the problem.

3. THE MATHEMATICAL STATE OF THE ART

The central mathematical question that arises in conjunction with singular perturbation problems is the following: Does each member of the family of boundary value problems A admit an asymptotic description of the form B? An epilogue to this question also asks: Can spurious asymptotic descriptions of the form B be distinguished from the correct one(s)?

Here we shall characterize the state of the art using two pairs of boundary value problems. The first member of each pair seems to typify the strongest theorems currently available.

Ia. Let

$$(3.1) \qquad \epsilon(u_{xx} + u_{yy}) - u_x = 0$$

in R, with $0 < \epsilon \ll 1$ and with u given on Γ. R is a region (such as that of Fig. 4) whose closed boundary Γ is nowhere tangent to a line $y = $ const. It has been proved [1] that u admits the asymptotic expansion

$$(3.2) \qquad u(x, y, \epsilon) \sim \sum_{n=0}^{\infty} u^{(n)}(x, y)\epsilon^n + \sum_{n=0}^{\infty} \epsilon^n U^{(n)} \left(\frac{x - x_1(y)}{\epsilon}, y \right),$$

and this series can be constructed uniquely. The function $x_1(y)$ is indicated in Fig. 4. That figure also distinguishes the region in which the second series contributes appreciably to the description from that in which it does not.

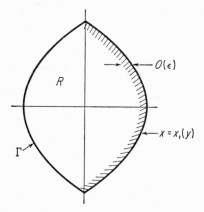

FIG. 4. Domain of Problem (Ia). Note that Γ is nowhere tangent to a line $y = $ const, and that the steep part of the structure is confined to the cross-hatched region.

Ib. Let Eq. (3.1) hold in R' a region that is exterior to the closed curve Γ' of Fig. 5 but between the vertical lines $(x = \pm L)$ of that figure. The function u is prescribed on $x = \pm L$ (without depend-

ence on ϵ) and on Γ'. Correct and convincing heuristic arguments
lead to the conclusion that the description of $u(x, y, \epsilon)$ is

$$u \sim \Sigma \, v^{(n)}(x, y)\epsilon^n = u_I \text{ in I,}$$

$$u \sim \Sigma \, w^{(n)} \left(\frac{r-1}{\epsilon}, \theta \right) \epsilon^n + u_I \text{ in II,}$$

$$u \sim \Sigma \, z^{(n)} \left(\frac{r-1}{\epsilon^{2/3}}, \frac{\frac{\pi}{2} - \theta}{\epsilon^{1/3}} \right) \epsilon^{n/3} + u_I \text{ in III,}$$

$$u \sim S[(\epsilon x)^{1/2}, y] = u_{IV} \text{ in IV,}$$

$$u \sim R(x, y) = u_V \text{ in V,}$$

$$u \sim \Sigma \, t^{(n)} \left(\frac{L-x}{\epsilon}, y \right) \epsilon^n + u_I \text{ in VI,}$$

and

$$u \sim u_V + \Sigma \, \epsilon^n Q^{(n)} \left(\frac{L-x}{\epsilon}, y \right) \text{ in VII.}$$

The series forms are avoided in the foregoing descriptions of u_{IV}
and u_V because we do not really know how the higher-order terms

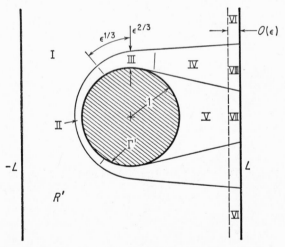

FIG. 5. The domain of Problem (Ib) with the multiplicity
of regions in which different characterizing behaviors of
the solution prevail.

look. For most purposes, in fact, the description of the solution in IV, V is best obtained by methods other than those of singular perturbation theory. The description in each of the regions I through VII has an overlapping region of validity with each contiguous region, so the composite description unambiguously accounts for $u(x, y, \epsilon)$ throughout R'.

This pair of examples illustrates that the answer to the mathematical question with which this section opened is well in hand for boundary value problems involving

$$(3.3) \qquad \epsilon L_1(u) + L_2(u) = w,$$

where L_1 is a second-order elliptic operator and L_2 is a first-order operator, but only if the domain is of a suitably restricted type (no boundary segment tangent to a characteristic of L_2). Not surprisingly, the level of intricacy that can be handled readily on a heuristic basis is much more extensive.

It is also not surprising that a heuristic approach is really very satisfactory for this class of linear boundary value problems.

In contrast to the foregoing, there seem to be no theorems that cover corresponding questions involving seriously nonlinear partial differential equations. Many such nonlinear problems can be handled heuristically, but we can give them no comprehensive characterization (except to say that they are the "easy ones"); they are illustrated extensively in the items listed in the References.

The second pair of examples is chosen from the realm of nonlinear ordinary differential equations.

IIa. Let

$$(3.4) \qquad \epsilon w_{xx} + a(w, x)w_x + b(w, x) = 0$$

in

$$A < x < B \quad \text{with} \quad 0 < \epsilon \ll 1,$$

with

$$a > 0,$$

and with

$$w(A) = p, \qquad w(B) = q.$$

It has been proved [2] that a solution $w(x, \epsilon)$ has an asymptotic expansion in powers of ϵ whose coefficients are functions of x, of $(x - A)\epsilon^{-1/2}$ and of $(x - B)\epsilon^{-1/2}$.†

A particular example of this is

$$(3.5) \qquad \epsilon u_{xx}(x, \epsilon) + 2b(1 - x^2)u(x, \epsilon) + u^2(x, \epsilon) = 1$$

in $|x| < 1$ with $u(1) = u(-1) = 0$, and the dominant terms of the description of a solution of this problem are

$$(3.6) \quad u(x, \epsilon) \sim u_1(x, \epsilon) = -b(1 - x^2) - \sqrt{b^2(1 - x^2)^2 + 1}$$
$$+ \frac{12e^{p_1}}{(1 + e^{p_1})^2} + \frac{12e^{p_2}}{(1 + e^{p_2})^2},$$

where

$$p_1, p_2 = \sqrt{\frac{2}{\epsilon}}(1 \pm x) + 2ln(\sqrt{2} + \sqrt{3}).$$

This function is plotted as the solid curve of Fig. 6 (for $b = 0$, the case for which a closed form exact solution can be constructed for comparison purposes).

IIb. This question asks whether other solutions of the boundary value problem associated with Eq. (3.4) exist and can be constructed by singular perturbation methods. The answer is yes, since this problem admits several solutions, many of which differ from each other at (say) $x = 1$ both in value and in slope by amounts of order $e^{-1/\sqrt{\epsilon}}$. Furthermore, it is very disconcerting to note that it is possible [3], by using singular perturbation methods, to construct spurious solutions of this boundary value problem. The asymptotic expansions of the boundary values and slopes of these solutions are identical with those of correct solutions. That is, true solutions differ at $x = \pm1$ from the apparent solutions by exponentially small discrepancies and there is *no way (known to me) within the framework of singular perturbation theory* by which the lack of validity can be detected. The dotted curve of Fig. 6 describes

---

† The translation I have of the theorem quoted from [2] must be in error. It is clear that counterexamples are readily found unless $a(w, A) = a(w, B) = 0$. However, this observation merely serves to reemphasize the subtleties which characterize the dividing line between that which has been proved and that which has not.

such a spurious result. It differs from u_1 by $12e^\zeta/(1 + e^\zeta)^2$, where $\zeta = (x - x_0)/\sqrt{\epsilon}$ and x_0 is any number $-1 < x_0 < 1$. Since, for $b = 0$, the boundary value problem can be treated exactly, each of these statements is readily verified.

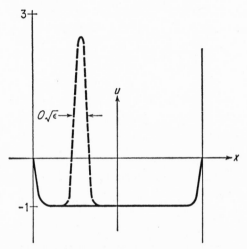

Fig. 6. A correct solution (solid wave) of Problem (IIa) and a spurious solution (dotted) of that problem.

Thus, the theorem exemplified under IIa is of limited value since, although it guarantees the existence of a solution of the prescribed form, it cannot provide protection against the construction of superficially acceptable but spurious solutions.

It is a correct conclusion that the analysis of an intricate and challenging problem of singular perturbation type will rarely be supported by theorems that guarantee existence, uniqueness, and/ or validity of description. But one should *not* conclude that intricate and challenging problems cannot be treated successfully; such methods are among the most successful tools for the treatment of important, nonlinear scientific questions. In particular, one's inalienable right to think while using any technique provides a very real degree of protection against the acceptance of the sorts of spurious constructions referred to above. The first conclusion implies no criticism of those who pursue such theorems; it is notoriously difficult to prove things about nonlinear boundary value problems, and the singular perturbation problems encountered in engineering and science are not heavily populated with the simpler of these boundary value problems. As we shall see in the subsequent sections, however, there are instances in which proofs of this sort would be of great value.

4. THE MID-LATITUDE OCEAN CIRCULATION

Observation of the motions in both the north Atlantic and north Pacific Oceans (together with lots of hindsight) suggest that it may be possible to understand the time-averaged large-scale features of these motions by treating the dynamics as though the fluid were confined by impermeable, but slippery, boundaries at 15° and 55° latitude. Such motions seem to be driven primarily by the time-averaged wind stresses whose (idealized) distribution is indicated in Fig. 7. The basin indicated schematically in that

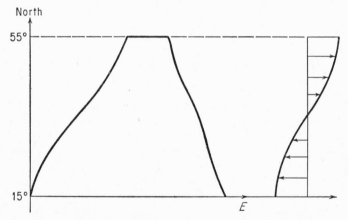

FIG. 7. Basin geometry and wind stress distribution for ocean circulation problem.

figure has an exceedingly irregular bottom topography, and the density distribution with depth that is typical of most of the basin is indicated in Fig. 8. There is no clear understanding of the way in which this density variation and the accompanying temperature and salinity variation are maintained; nor is there a better understanding of the importance to the dynamics of the water circulation of this density variation and of the topographical vagaries of the bottom.

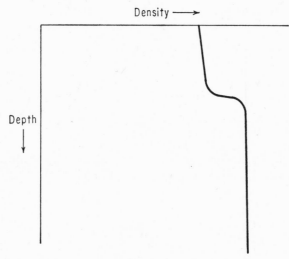

FIG. 8. Typical (smoothed) mid-latitude density distribution.

Furthermore, it is not yet possible to decide whether the time dependent fluctuations of velocity have a macroscopic effect beyond those that are adequately modeled by a large equivalent viscosity. Many other items that are potentially important also are not understood, but there is no further need for enumeration. What should already be clear is that one cannot initiate attempts to explain the size, speed, location, and meandering of the Gulf Stream and/or the Kuroshio by invoking a theory that takes into account the details of all of these complicating features. Furthermore, such a theory is not a primary objective. What *is* needed is an identification of those mechanisms, major features of the medium and its environment, and external inputs that might account for the kind of phenomenon that is known to occur. The

simplest plausible model based on tentative identification of these
items can be studied, and the phenomenon that would occur in the
situation to which the model really applies could be deduced.
Comparison of those deductions with observation then leads to a
revised model and the process continues.

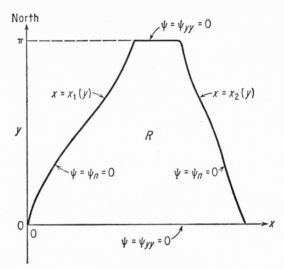

FIG. 9. Domain and boundary conditions for ocean current
problem.

In the model that has received the most attention over the past
20 years, considerations of mass and momentum conservation lead
to the boundary value problem

$$(4.1) \qquad \epsilon \Delta \Delta \psi + \alpha(\psi_y \Delta \psi_x - \psi_x \Delta \psi_y) - \psi_x = \sin y$$

in R (see Fig. 9) with boundary conditions indicated in that figure.
In this model, ϵ and α are both very small numbers. The curves,
$\psi = $ const, are the depth-averaged particle paths of the fluid.

The four individual terms arise, respectively, from considerations
of frictional resistance, the acceleration of the fluid relative to the
earth, the change with latitude of the Coriolis acceleration associated with the earth's rotation, and the stress exerted by the
wind. The value of ϵ that "parameterizes" in a very crude way the

complications of turbulent momentum transfer is not known. Rough estimates can be inferred by using observational information about other phenomena in the same medium. The simplest geometry that makes any sense at all has been adopted.

It should be clear that this model will prove to be a viable one *only* if the density variations, topographical variations, input variations, and other unmentioned complications all happen to be relatively unimportant in the determination of the major features of the phenomenon. Thus, the extent to which this model will ultimately aid our understanding of ocean currents cannot yet be decided. But it is certainly a proper formulation of the steady state dynamical problem in which a layer of homogeneous viscous fluid between slippery spherical caps in a rotating system is forced into motion by a body force distributed uniformly over the depth; this is an observation that will have significance later in the discussion.

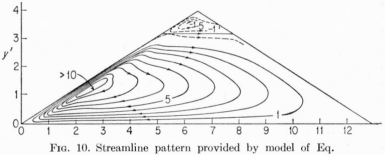

FIG. 10. Streamline pattern provided by model of Eq. (4.1) with $\alpha = 0$.

In an earlier version of this model, the second term was ignored, and the linear singular perturbation problem so posed was readily treated and (for the basin geometry shown) led to the flow pattern of Fig. 10. Its analytical description is given by

$$\psi = \psi \sim \psi^{(1)}(x, y, \epsilon)$$

$$= \sin y \left[x_2(y) - x - \{x_2(y) - x_1(y)\} \right.$$

$$\left. e^{-Q} \left\{ \cos (Q\sqrt{3}) + \frac{1}{\sqrt{3}} \sin (Q\sqrt{3}) \right\} \right],$$

where

$$Q = [x - x_1(y)]/2(\epsilon\mu)^{1/3}, \qquad \mu = \{1 + [x_1'(y)]^2\}^2.$$

However, the value of ϵ that provided a stream width compara-
ble to that of the real current was larger by a factor 10^3 than can
be readily accepted, although this statement will be reopened for
discussion later. Even with that value, the geometry of the pattern
is unsatisfactory (see Fig. 11), and it is not hard to verify that the
second term of Eq. (4.1), calculated from the tentatively adopted
description $\psi^{(1)}$ cannot justifiably be ignored. A boundary layer
technique was not the only one by which this problem could be
(and was) solved. However, the readily interpreted and readily
extended description to which this method gave rise displayed
more clearly than any other description (of the same function) the
fact that neither the postulation of alternative geometric details
nor acceptable modifications in the guess-estimated value of ϵ
could remove the inadequacy of the results.

These observations led to the study of the more elaborate model
involving all of Eq. (4.1), and had superficial and short-lived
success. It was possible, despite the nasty nonlinearity, to con-
truct, analytically, singular perturbation descriptions of ψ in
$y < \pi/2$ that implied the velocity profile of Fig. 12. The description is again of boundary layer type,† but, unfortunately, when
$\epsilon^{1/3}$ is small compared to $\alpha^{1/2}$, the solution cannot be continued
into the upper half of the basin. It is also interesting to note that
attempts to integrate this problem numerically have proved
fruitless except for values of ϵ that are very much larger than those
ordinarily regarded as realistic.

The physical complication is readily exposed. To follow the
argument, suppose tentatively that the velocity pattern is like
that of Fig. 12 along the entire western side of the basin. As a
particle on any streamline moves from its position at time t (say *
on Fig. 10) along the streamline and returns at a later time t to *,

---

† A synonym for the results of informal singular perturbation methods when
the small scale features of the phenomenon are confined to the neighborhood
of a boundary segment.

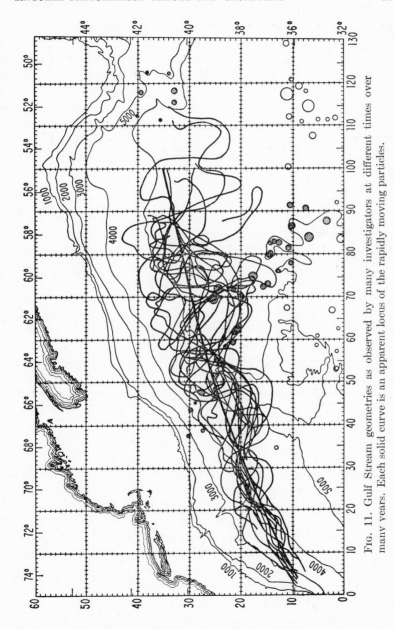

FIG. 11. Gulf Stream geometries as observed by many investigators at different times over many years. Each solid curve is an apparent locus of the rapidly moving particles.

it accumulates vorticity (vorticity = spin = $\Delta\psi$) from the wind stress, and this input is everywhere positive.

The convection of vorticity and the consequences of the Coriolis contribution [terms 2 and 3 of Eq. (4.1)] do not include any possibility for the transfer of any of that vorticity to another particle. Since our model requires a time-independent phenomenon, the diffusive term (i.e., the friction term) *must* provide for the removal of all the added vorticity by the time the particle has returned to *. But if the structure shown in Fig. 12 were valid in

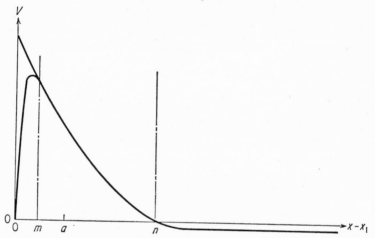

Fig. 12. Northward velocity profile. Frictional mechanisms are important in 0 m, the current mn is controlled by a balance of Coriolis acceleration, and "$\vec{V}\cdot\text{grad }\vec{V}$."

$0 < y < \pi$, the diffusive term is of significant size only in a region whose thickness is governed by the size of ϵ as indicated in that figure. That is, if ϵ is too small, a particle passing to the right of the streamline through a (in Fig. 12) on its northward journey is not subjected appreciably to the diffusive process, and it cannot return to * with the same vorticity it had on the previous trip. Thus, somehow, *the particle paths must be more tortuous than those shown* or *there is no solution to this problem with $\epsilon^{1/3} \ll \alpha^{1/2}$*. It is hard to accept the latter possibility, because the problem posed *is* a legitimate model of a conceptually acceptable fluid mechanics experiment in which one would expect a steady flow to exist even though it might be unstable. Thus, if we tentatively reject the latter possibility, we can argue for a tortuous structure by noting that such a flow almost certainly *would* be dynamically unstable. Such an instability might easily foil the numerical efforts, and it would imply physically that the "steady" flow would break down into something "turbulent" or "undulatory" or something similar. If one looks at the highly irregular structure so implied (and so characteristic of the real ocean) on a really long time average, the "steady" picture so generated involves a wide current with big enough fluctuations in its "microstructure" that a much larger value of ϵ could be used to model the accompanying mixing process in $y > \pi/2$. This almost puts us back to the early, larger ϵ theory, provided that the bottom topography or the presence of colder heavier water north of the stream can help to explain the fact that its mean trajectory continues to head out to sea from exactly the place from which the small ϵ theory cannot be continued.

It is clear that the foregoing phase of the study of the mid-latitude circulation has again gained much from the use of singular perturbation ideas. The recognition that a narrow current cannot proceed northward beyond the point where the driving term ($\sin y$) has its maximum is not easily made by alternative means. Furthermore, the recognition that the flow must have a multiscale character leads directly to the alternatives underlined above. Finally, if there is a solution associated with this model, one must look beyond the boundary layer structures that have characterized the descriptions that have arisen in the earlier studies. It is highly probable that any satisfactory description of a solution will be multiscaled in the sense of (but more complicated than) the undulatory (one-dimensional) phenomena of Figs. 2 and 3.

No successful attempts have been made to cope with this problem yet, and, in fact, this is one of those relatively rare studies of an individual phenomenon to which a firm answer to the question, "Does the solution exist and does it have the form X?" would be extremely helpful. In fact, a firm answer could be crucial in the choice of the next line of attack. The papers that are collected in [4] give a comprehensive account of the chronological evolution of the model so briefly reviewed in this section. Any feeling that the present treatment has been too brief is supported by the observation that 161 pages are devoted to this topic in [4].

5. HURRICANES

Each year several large cyclonic storms are initiated over the oceans, usually in the general vicinity of 15° latitude. These storms are characterized by maximum wind speeds that occasionally may be as great as 200 miles per hour. The area over which the organized motion occurs is of the order of 1000 miles in diameter, and the depth of the portion of the atmosphere that is significantly involved in the motion is a very few miles. The general structure seems to be consistent with the idealization shown schematically in Fig. 13. It consists of four different pieces. In region II the

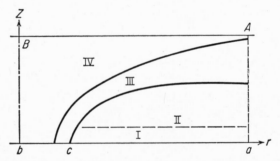

Fig. 13. A schematic view of the four regions associated with a mature hurricane.

fluid is swirling rapidly; it also is settling very slowly, but the radial velocity is extremely small. Frictional effects are important in region I; there, in fact, the frictionally reduced peripheral velocity implies that the centripetal acceleration in that region cannot balance the pressure gradient implied by the larger swirl in II, and the excess pressure gradient drives a radial influx of fluid. This fluid moves upward in the annular region III, where, at any given radial position, it has much less angular momentum per unit mass than the fluid in II at the same radius, because much of the angular momentum of this fluid was negated by the frictional forces exerted across the water surface as the fluid moved radially inward in I. The fluid in III also is less dense than nearby fluid in II for reasons that will be discussed soon. The fluid in the central region (or core) is relatively motionless and is very warm and dry. These statements about density imply that a column of air that is closer to the center than a second column has less total weight than the second one; this is a necessary condition for the existence of the horizontal pressure gradient that balances the centripetal acceleration field in region II.

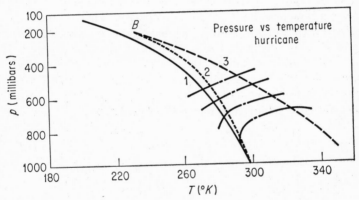

Fig. 14. Typical pressure-temperature plots for West Indian hurricanes. Indicated are (1) a typical ambient profile; (2) the moist adiabat based on ground level ambient conditions; (3) the dry adiabat recompression from B of the air, which underwent a moist adiabatic expansion along (2) to B.

The thermodynamic situation is depicted in Fig. 14. There, in curve 1, the pressure and temperature of the ambient air (that at a of Fig. 13) are related to each other. This air is very moist and quite warm, because the ocean has been heating it throughout the tropical summer. When air from the bottom of this region is moved inward and then into the updraft region, its moisture condenses (the moisture in this air reaches the saturation level early on this journey as the pressure and temperature drop almost adiabatically), and the released heat of vaporization is retained by the rising air. The thermodynamic path for this bit of history is also denoted in Fig. 14 on curve 2, and one sees that, at a given altitude, the density must be smaller in region III than it is in region II at $r = a$. This rising air also entrains small amounts of air from the core and induces a slow circulation in that core. Probably, in fact, it is the gradual drying of this air in the upward part of that circulation followed by the adiabatic compression of the dry air in the return motion, during the formative stages of the storm, that accounts for the state of the gas in the core. If this hypothesis is used, the third curve of Fig. 14 depicts the state of this gas that has the least possible density. That is, the state on curve 3 is that of gas, all of which has "gone up" curve 2 to B and and then down curve 3. Using these loci of thermodynamic states, one can readily calculate the largest pressure discrepancy at ground level that is available to balance the centripetal acceleration. It is quite evident observationally that velocity distributions actually achieved in region II of real storms are sandwiched between one extreme, $V \sim 1/r$ and a second, $V \sim 1/\sqrt{r}$. The maximum speed that can be achieved with either of these distributions is independent of the lateral size of the storm and is such, for typical tropical atmospheres, that speeds in excess of 200 miles an hour are consistent with this model. This maximum speed consideration is crucial if the model of the storm is to be viable without reliance on very rapid heat and moisture transfer from the ocean surface to the radially moving fluid in the boundary layer.

Only indirectly have singular perturbation ideas played any role in the foregoing aspects of this model. However, there are two vital questions that must be answered quantitatively before the validity of the model can be assessed, and these will necessarily rely for tractability on boundary layer ideas.

The first question has to do with region I. It asks: Is there a family of swirl distributions beneath which a thin layer of fluid can move radially inward in such a way that, over $c < r < a$, the vertical velocity at the upper edge of this layer is almost everywhere downward? The fact that this question can be so phrased implies a conviction on the questioner's part that individual facets of the problem can be isolated, and this, of course, is a central aspect of singular perturbation techniques. The smaller length scale l that characterizes the height of this layer is

$$l = (\nu/2\Omega)^{1/2},$$

where ν is an effective turbulent diffusivity and Ω is the angular velocity of the earth. The equations to be dealt with are

(5.1) $\quad \phi\phi_x + w\phi_z + (\Psi^2 - \psi^2 - \phi^2)/2x - (\psi - \Psi) = \phi_{zz},$

(5.2) $\quad\quad\quad\quad \phi\psi_x + w\psi_z + \phi = \psi_{zz},$

(5.3) $\quad\quad\quad\quad\quad\quad \phi_x + w_z = 0,$

where z = vertical coordinate $\div\ l$, x is proportional to the square of r (Fig. 14), and Ψ is a known function of x that characterizes the state of affairs in region II. We include these equations only to indicate the level of complication; it will be difficult to find a reliable description of flow fields associated with the question raised above if, indeed, they do exist.

Without some suitable approximation to such a solution, one cannot estimate reliably the radial flux of fluid or the flux of heat and moisture from the ocean to that fluid. Preliminary estimates do imply that the heat and moisture transfer to the boundary layer air cannot be large enough to play a crucial role in the phenomenon, but only extremely rough estimates of the mass flux have been made.

Another question that is equally crucial to the assessment of validity has to do with the coupling between regions II and III. It asks: Can the fluid in region II flow past that in region III without a very significant transfer of angular momentum from one body of fluid to the other?

If the fluid in II near the "interface" of II and III were not very different in density from that of III near the "interface," the

answer to the question would be no. By using boundary layer
ideas again, it is relatively easy to calculate that the loss of
momentum from II to III in a configuration of this sort would
render it impossible to maintain swirl distributions of the sort
described. In fact, very little swirl could continue to occur any-
where except in the immediate vicinity of the updraft. It is also
known, however, that there are many configurations in which two
contiguous rotating bodies of fluid, having significantly different
densities, can display motions which are *not* strongly coupled.
No such statement can yet be made with assurance for the configuration of Fig. 13; the analysis that will underlie such a statement, in fact, will certainly involve the description of a *very*
multiscaled situation and will certainly rely on boundary layer
ideas.

6. A FINAL REMARK

There is one important characterization of singular perturbation
problems that was not emphasized in Sections 2 and 3. It did
emerge in Sections 4 and 5 that, typically, the description of the
functional dependence of the dependent variables (say ψ_i) on a
small-scale independent variable (say x/ϵ) is studied without
interference from the complications arising from the dependence
of ψ_i on the other variable(s) (say x). Often this decomposition
of the problem provides an enormous simplification, not only in
the analysis but especially in the description of the answer. This
simplicity of description, and therefore of interpretability, is probably the most important single aspect of this family of techniques;
the assessment of the validity of a model and the clarity with
which one can interpret what he has accomplished is certainly
crucial to the value (especially to the reader) of the end product of
the study.

All of these claims become much more visible when one follows
detailed applications of the ideas to specific problems. The reader
is urged to negate the inadequacies of this brief and fragmented
review by following the analysis of some of the many fascinating

illustrations listed in the bibliography. Note, in particular, that item [8] itself contains 183 bibliographical items.

REFERENCES

1. Eckhus, W. and E. M. DeJager, "Asymptotic solutions of singular perturbation problems for linear differential equations of elliptic type," *Arch. Rat'l Mech. Anal.*, **23** (1966), 26.

2. Vasil'eva, A. B., "Asymptotic behaviour of solutions of the problems for ordinary non-linear differential equations with a small parameter multiplying the highest derivatives," *Uspehi Mat. Nauk*, **18** (1963), No. 3 (111), 15–86. (Translated in *Russian Math. Surveys*, **18**, No. 3, 13).

3. Carrier, G. F. and C. E. Pearson, *Ordinary Differential Equations*. Waltham, Mass.: Blaisdell, 1968.

4. Robinson, A. R., *The Wind Driven Ocean Circulation*. Waltham, Mass.: Blaisdell, 1963.

5. Feshchenko, S. F., N. I. Shkil, and L. D. Nikolenko, *Asymptotic Methods in the Theory of Linear Differential Equations*. New York: American Elsevier, 1967.

6. Kaplun, S., *Fluid Mechanics and Singular Perturbations*, P. A. Lagenstrom, L. N. Howard, and C. S. Liu, eds. New York: Academic Press, 1967.

7. Wasow, W., *Asymptotic Expansions for Ordinary Differential Equations*. New York: John Wiley—Interscience, 1965.

8. O'Malley, R. E., Jr., "Topics in singular perturbations," *Advances in Mathematics*, **2**, 365–470. New York: Academic Press, 1968.

9. Fraenkel, L. E., "On the method of matched asymptotic expansions: Part I, a matching principle; Part II, some applications of the composite series; Part III, two boundary-value problems," *Proc. Cambridge Philosophical Soc.*, **65** (1969), 209–231, 233–261, 263–284.

10. Exkhaus, W. E., "On the Foundations of the Method of Matched Asymptotic Approximations," Mathematisch Instituut der Technische Hogeschool Delft, Nederland (July 1968).

11. Cole, J., *Perturbation Methods in Applied Mathematics*. Waltham, Massachusetts: Ginn-Blaisdell, 1968.

12. Van Dyke, M., *Perturbation Methods in Fluid Mechanics*. New York: Academic Press, 1964.

13. Greenspan, H. P., *The Theory of Rotating Fluids*. Cambridge: Cambridge University Press, 1968.

NONLINEAR DIFFUSION PROBLEMS

Hirsh Cohen

1. INTRODUCTION

Classical mathematical physical descriptions of linear phenomena have a convenient proto-typification into elliptic, parabolic, and hyperbolic partial differential equations. The physical notions of static equilibrium, diffusive decay, and wave motions with conserved energy are all made very explicitly understandable with reference to Laplace's equation, the heat equation, or the wave equation. This is true both for the kinds of force, energy, momentum, or material balancing that go into the equation derivations as well as for the solutions. Linearity allows for superposition of solutions, and thus all of the familiar elements of discrete and continuous compositions are available. In direct contrast, when nonlinearities occur strongly in the equation, the kinds of phenomena that appear are not so familiar; they are, in fact, only beginning to be classified or even recognized. It is just this aspect of nonlinear equations that we would like to examine in this article. We shall try to do this by looking at several special examples of nonlinear diffusion equations.

The expectations of linearity have been so pervasive that we expect diffusion equations to indeed describe "diffusive" behavior —as time increases, initial amplitudes and other features are smeared out. Therefore, when nonlinearity provides traveling waves or frontal motions of other kinds, we are surprised. We would like, in fact, to give some idea, through these examples, of how rich the phenomena can be that even simple nonlinearities provide, and how surprising they may be.

The discussion that follows will focus on two kinds of problems in which finite velocity fronts or pulse, wavelike motions occur. For this to happen, there must be energy released or inserted into the field so that the normal diffusionlike decay is balanced out. We shall be rather explicit in describing physical aspects of the examples. In the first of these, the nonlinearity is present as an added source in equations that describe electrical impulse conduction along nerve membranes, flame front propagation, and in a model of the advance of mutations. In the second, we describe a classical model, the Stefan problem, which represents the propagation of a phase transition. Applications in superconductivity, geophysics, soldering, and melting and freezing of ice are discussed. In this case, too, energy is extracted from the material (the latent heat) as a function of the solution.

Although these two examples form the main content, there are a number of other examples that are at least as important. In fluid dynamics, the equations of viscous flow describe the diffusion of vorticity in the presence of convective forces. The balancing of these forces leads to a large number of interesting phenomena. The well-known concept of the boundary layer and the methodology of singular perturbations came originally out of these equations.

In another class of problems, the coefficients in the diffusion equation depend on the solution function. As one can see from some of the numerically calculated examples in the book by Crank, *The Mathematics of Diffusion* (Oxford Press, London, 1956), a large class of such nonlinear diffusion coefficients shows only quantitative changes from the related linear problem. For example, for the problem

$$\frac{\partial}{\partial x}\left(\left(1 + \alpha\,\frac{c}{c_0}\right)\frac{\partial c}{\partial x}\right) = \frac{\partial c}{\partial t}, \qquad \begin{array}{lll} c = 0, & x = 0, & t > 0 \\ c = c_0, & x > 0, & t = 0 \end{array}$$

when $\alpha = 10$, the results are qualitatively similar to the results for $\alpha = 0$. However, for the case in which the diffusion coefficient is $\alpha c/c_0$ instead of $1 + \alpha(c/c_0)$, i.e., the diffusion coefficient can be zero (the boundary value), there can be a finite velocity for the value $c = 0$ as it travels into the field. Since in linear diffusion all signals propagate with infinite velocity to all points of the field, this is a rather striking result. In this case, the value $c = 0$ propagates with velocity $1.31\alpha/x$ (Crank, p. 165). Julian Cole of the University of California at Los Angeles has pointed out that there are several other examples of nonlinear diffusion coefficients with which finite velocity frontal propagation may be associated. We shall not discuss these further, although they are apparently somewhat like the problems we do discuss in detail.

In another large class of problems, the flux of charged bodies is determined by the combination of a diffusion potential and another potential force field. For example, at a semiconductor junction, the distribution of electrons and holes is governed by nonlinear diffusion equations of the form

$$\frac{\partial n}{\partial t} = -\frac{\partial J_n}{\partial x} - R_n(n, p),$$

$$\frac{\partial p}{\partial t} = \frac{\partial J_p}{\partial x} - R_p(n, p),$$

$$J_p = q\mu_p\left(p\,\frac{\partial \psi}{\partial x} + \alpha_p\,\frac{\partial p}{\partial x}\right),$$

$$J_n = q\mu_n\left(n\,\frac{\partial \psi}{\partial x} - \alpha_n\,\frac{\partial n}{\partial x}\right).$$

In these equations, n and p are the concentrations of electrons and holes, J_n and J_p are their current fluxes, R_n and R_p are recombination functions, q is the electric charge constant, and μ_p, μ_n, α_p, and α_n are mobility and diffusion functions, respectively. The function ψ, the electric field potential, satisfies Poisson's equation

$$\frac{\partial^2 \psi}{\partial x^2} = -q(p - n + D)$$

in which D is a fixed charge and q is a measure of the electrolytic charge constant.

A very similar set of equations governs the behavior of different species of ions in solution and is an important model of the physical chemistry of the solutions.

In both cases, although there are special conditions in which particular nonlinear solutions may be analyzed, one understands well that there is a rich amount of nonlinear behavior—for example, large changes occurring in small regions in transistors give rise to the high switching speeds in these devices—that is observed experimentally but is hard to get at mathematically.

2. ELECTRICAL CONDUCTION IN NERVE AXON

a. HODGKIN-HUXLEY THEORY. The voltage change $v(x, t)$, which travels along a nerve axon from the nerve cell body to the termini of the axon, has the form of a sharp-front pulse and propagates with constant velocity. This fixed shape, constant velocity pulse travels from one end of the axon to the other; information is carried not in the amplitude or shape of a single pulse but, rather, digitally, in the frequency of arrival of pulses. Because the axon has a diameter very large compared to the thickness of the membrane that surrounds it, it is possible to represent the propagation as one-dimensional along the axial direction. And because magnetic effects are negligible in nerve tissue, the equation governing the voltage across the membrane is

$$(2.1) \qquad c\,\frac{\partial v}{\partial t} + I_i(v) = \frac{a}{2R_i}\,\frac{\partial^2 v}{\partial x^2}.$$

This corresponds to an equivalent circuit for the membrane of the form shown in Fig. 1. The nerve cell is shown in Fig. 2. This is a current density equation. The currents referred to that support the voltages are flows of ions across the membrane into and out of the axon. The term $c(\partial v/\partial t)$ is the current density that flows through the fixed capacitance c; $I_i(v)$ is the current density across the resistive element of the membrane; the term $a/2R_i\ \partial^2 v/\partial x^2$ is

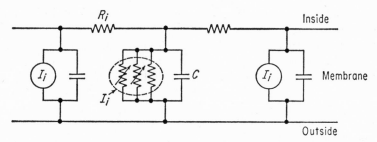

FIG. 1. Equivalent circuit for nerve membrane. In the passive state $I_i = \dfrac{v}{R_s}$; in the active state I_i is as shown in the dotted circle.

the sum of these that flows along the axon. a is the axon radius, and R_i is the specific resistance of the electrolyte inside the axon.

If the cross-membrane ionic current term were linear, v/R_s say, then the response to a stimulus at one end of the cell would decay very rapidly. In the squid, the giant axon may be 20 cm long; values for R_i, R_s, c, and a are such that the stimulus amplitude would decay by one-third in one cm. Thus, faithful conduction to the end of the cell could not take place.

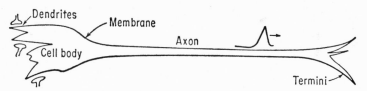

FIG. 2. A nerve cell.

A dependence of I_i on $v(x, t)$ is required. Its actual form has been found to be quite complex; it has, however, been determined for several kinds of nerve membrane. Notably, Hodgkin and Huxley [1] have found, for the squid axon, that the cross-membrane current can be expressed in the following manner:

$$(2.2) \quad I_i = g_{\mathrm{Na}}(v)(v - \bar{v}_{\mathrm{Na}}) + g_{\mathrm{K}}(v)(v - \bar{v}_{\mathrm{K}}) + \bar{g}_l(v - \bar{v}_l).$$

Barred quantities are constants; $\bar{v}_{\mathrm{Na}}$, $\bar{v}_{\mathrm{K}}$, and $\bar{v}_l$ are values of equilibrium voltages that are dependent on the ratios of ionic concentrations of sodium, potassium, and other ions inside and outside the membrane.† g_{Na} and g_K are variable conductivities related, respectively and separately, to sodium ion and potassium ion flow. $\bar{g}_l$ is the constant conductivity associated with chloride and the remaining ions of the external and internal electrolytic baths.

To describe the nonlinear conductivities, g_{Na} and g_K, Hodgkin and Huxley measured ionic currents carefully and assigned three new variables: m and h describe, in turn, the way in which sodium ions begin to flow into the membrane and then the slowing down of that inflow; n describes the voltage-dependent time course of the potassium ions. The equations are nonlinear, first-order kinetic descriptions of the ion change:

$$(2.3) \qquad \frac{\partial m}{\partial t} = \frac{1}{\tau_m(v)} \, (m_\infty(v) - m),‡$$

$$(2.4) \qquad \frac{\partial h}{\partial t} = \frac{1}{\tau_h(v)} \, (h_\infty(v) - h),$$

$$(2.5) \qquad \frac{\partial n}{\partial t} = \frac{1}{\tau_n(v)} \, (n_\infty(v) - n).$$

The τ's and m_∞, n_∞, and h_∞ are measurable functions and must, in fact, be measured for each kind of membrane. τ_m is much smaller than the other two "time-constant" functions, reflecting a very fast early sodium response as the voltage begins to increase.

---

† E.g.,

$$\bar{v}_{Na} = \frac{RT}{F} \ln \frac{c_{Na,0}}{c_{Na,i}}.$$

RT/F is an electrolytic constant, and $c_{Na,0}$ and $c_{Na,i}$ refer to concentrations of sodium outside and inside the membrane.

‡ As an example,

$$\frac{1}{\tau_m(v)} = \frac{0.1(25 - v)}{\left[\exp\left(\dfrac{25 - v}{10} \right) - 1 \right]} + 4 \exp\left(-\frac{v}{18} \right),$$

$$m_\infty(v) = \frac{0.1(25 - v)\tau_m}{\left[\exp\left(\dfrac{25 - v}{10} \right) - 1 \right]}.$$

These five equations form a system that should describe the propagation of voltage impulses along the axon membrane. Let us look explicitly at what we shall ask of solutions:

1. If the dependence of v on x is removed, an impulse can be stimulated simultaneously at every point. This is called the "clamped" experiment. An impulse will arise only if the stimulating current or voltage reaches a threshold value. When this occurs, the voltage continues to rise without further input to a peak of about 100 millivolts and then to fall off to a small negative value before returning to the starting level.

   For the system of equations $\partial/\partial x = 0$, we must solve the fourth-order set of ordinary differential equations as an initial value problem.

2. Boundary conditions are given at a point on the axon, say $x = 0$, of the form $v(0, t) = v_0(t)$; at a point far down the axon, $x \to +\infty$, $v \to 0$. The boundary point function must again reach a threshold value, and a pulse similar to the "clamped" pulse now propagates with a constant velocity.

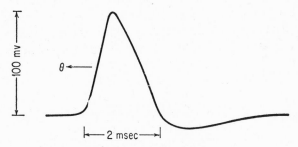

FIG. 3. A nerve voltage pulse.

3. We can also assume that there exists a steady propagating pulse, $v = v(x + \theta t)$, and seek solutions of the proper shape and velocity of propagation θ. Figure 3 shows what we expect the pulse to be like, and Fig. 4 shows its relationship to m, n, and h.

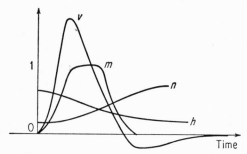

Fig. 4. The relationship between the voltage v and the three ionic variables m, n, and h.

4. The threshold relation—the length of time a stimulus must be maintained to create a pulse—should be calculable.
5. The real function of a nerve axon is to conduct sequences of pulses along its length. The time interval between pulses or the frequency of pulse trains or the appearance of pulses in finite trains forms the information coding of the nerve system. The Hodgkin-Huxley equations should be able to produce such repeated pulse patterns.
6. Although the nerve membrane operates normally in particular concentrations of sodium, potassium, and other ions, it can be shown experimentally that adding other chemicals—drugs and poisons such as procaine or tetrodatoxin—alters the pulse shape considerably. Mathematical models of the chemical-electrical interaction should reflect this.

These are not the only demands on the model, but they are a sufficient list to show that there must exist a rather rich set of mathematical phenomena associated with the solutions. The difficulty is, however, that the special functions that characterize the nonlinearity—the τ's and m_∞, n_∞, h_∞ and the appearance of m^3 and n^4—do not offer much hope of useful quantitative analysis of the equations. A good deal of numerical work has been done, and we can report on this, but, before doing so, we shall try to show—through simpler examples—some of the qualitative properties of equations of this type.

b. SIMPLER EQUATIONS. In order to see what the non-linear diffusion process is like, let us consider the single equation

$$(2.6) \qquad v_t + f(v) = v_{xx}.$$

Equations of this type occur in flame propagation theory [2, 3] and in some simple theories of the propagation of advantageous mutations [4].

If we demand solutions that have a fixed shape and a constant propagation velocity, we can set

$$(2.7) \quad v = v(x + \theta t), \qquad x + \theta t = \xi; \qquad v = 0, \qquad \xi \to +\infty,$$

$$v = 1, \qquad \xi \to -\infty$$

and produce the ordinary differential equation

$$(2.8) \qquad v'' - \theta v' - f(v) = 0.\dagger$$

By considering the critical points in a v, v' phase, it is easy to analyze a number of particular nonlinearities $f(v)$ (see Table 1).

In case (e), as v^* approaches zero, it can be shown that $\theta \to 2$ [5]. This resolves the nonuniqueness of (a) and (g). Another way to resolve this problem has been studied by Kolmogoroff, et al. [6]. Their method is to return to the partial differential equation $v_t + f(v) = v_{xx}$, define a suitable initial value, and show that as $t \to \infty$, $v(x, t)$ goes over to the solution of

$$v'' - \theta v' - f(v) = 0$$

for which $\theta \equiv 2$.

A similar discussion has been given by Kanely [7] for cubic functions like case (f) in Table 1. The results of his analysis from the initial value problem are not so clear-cut.

All of these nonlinearities lead only to traveling fronts. The relationship to the Hodgkin-Huxley equations is seen if we arrange to hold n and h constant in that system. One can then calculate a traveling front solution. This can be experimentally verified by adding tetraethyl ammonium chloride to a squid axon bath (see calculations and analysis by FitzHugh [8] and Noble [9]). The ion outflow is blocked and a front travels along the axon.

---

† " ′ " means differentiation with respect to ξ.

Table 1

$f(v)$	Solution shape	θ
a. $v(1 - v)$		$\theta \geq 2$
b. $v(1 - v)^2$		θ unique
	No solutions of this type	
c. $v^2(1 - v)$		θ unique
d. $-v^2(1 - v)$		unique solution
e. $f(v) = g(v) = 0$ $1 \geq v \geq v^* > 0$ $v^* \geq v \geq 0$		
f. $v(1 - v)(a - v)$, $0 < v < 1$		$v = \dfrac{1}{1 + e^{-\xi/\sqrt{2}}}$; $\xi = x + \theta t$, $\theta = \sqrt{2}(\tfrac{1}{2} - a)$ $(a = 0, \text{ case d})$
g. $f(v) > 0$, $0 < v < 1$ $f(0) = f(1) = 0$, $f'(0) > 0$, $f'(1) < 0$		$\theta \geq 2|f'(1)|$

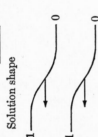

What is interesting is that at least in the calculations the velocity of propagation of the front is very close to that calculated for the pulse.

Returning now to the numerical solutions for the Hodgkin-Huxley equations, we show in Figs. 5 through 7 the kinds of results that may be obtained. For the space-clamped calculation, the numerical results give a pulse that very closely matches the experimental one. For the traveling wave solution, Huxley [10] calculated, by a numerical shooting method, that there was not only a pulse solution of the proper amplitude, shape, and velocity to match experimental results, but also two other solutions—a low-amplitude, slower-velocity pulse and an oscillation. Cooley and Dodge [11] have shown numerically that the initial value problem approaches the higher pulse and larger velocity as the pulse proceeds down the axon. They were also able to obtain low-amplitude, slow pulses but only for very brief times and distances along the axon when the normal ionic functions were used. When the axon is (mathematically) treated with drugs, the sodium and potassium conductivities are altered. This can be shown to have an effect on the amplitude and speed of pulse propagation that brings the large and small pulse closer together (Fig. 5). Threshold curves that exhibit how long a stimulating input must be maintained to set off a pulse can be found by calculation (Fig. 6). Finally, the amplitude-to-frequency transduction property of the axon and its equation can be computed (Fig. 7). One sees that as the current stimulus amplitude is increased, over a very short range there is a finite number of pulses set off—one, two, up to five or six, and then for higher values, an infinite train. Actually, experimentally, nerve fibers respond in a more complicated fashion—the range in stimulus for finite trains is broader, and the pulse interval decreases as the stimulus increases—but recently, Dodge and Hodgkin have shown that the Hodgkin-Huxley theory can accommodate these experimental results.†

One should bear in mind that these are all numerical results.

---

† Private communication.

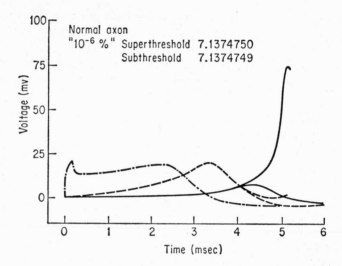

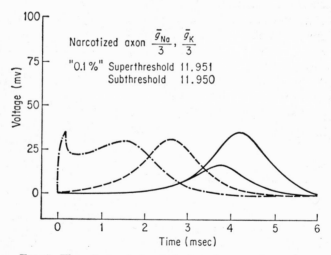

FIG. 5. The effects of chemical alterations on the nerve pulse: dashed-dot curves are at $x = 0$, dotted curves at $x = 2$ cm, and solid curves at $x = 5$ cm. Notice the relative persistence of the unstable (low amplitude pulse) in the "doped" axon.

Very little, if any, analytical work has been done on the mathematical mechanisms of these effects. The last one mentioned, for example, suggests that in a five-dimensional phase space (for the steady propagating pulses) the solution cycles around a limit cycle (or very near one) a small number of times and then leaves the

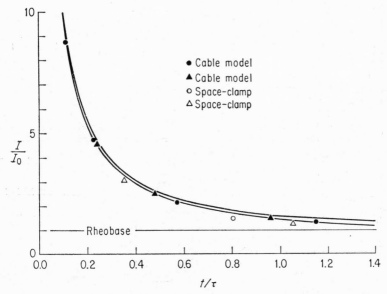

Fig. 6. Calculated threshold curves. The two sets represent different temperatures. I/I_0 is a normalized current stimulus amplitude, and t/τ is a normalized time.

limit cycle to return to an equilibrium point. The degree of variability available by alteration of the conductivities—the sensitivity to chemical change—has not been qualitatively explored. The stability of such traveling pulses has not been analyzed. There is a linear theory of these equations [1, 12, 13] from which one may be able analytically to produce the threshold function, but this has not been literally done either.

c. AN APPROXIMATE MODEL. In order to gain more insight into such mathematical mechanisms, one can look at an

approximation for the Hodgkin-Huxley theory. The equations
we will now discuss were first proposed by FitzHugh [14] and later
analyzed to some extent by Nagumo, et al. [15].

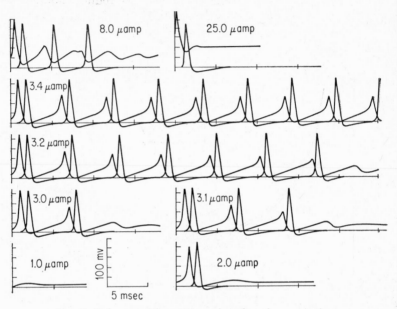

FIG. 7. Pulse repetition rate as a function of current stimulus.

Suppose we return to example (f) in Table 1 for the function
$f(v)$, the cubic

$$f(v) = v(1 - v)(a - v).$$

As we have seen, this gives a traveling front whose velocity depends
on the parameter a but whose shape is fixed. To produce a nerve-
like pulse, we need to bring the flat top of the front back down
below the zero level and then to zero again. Physically, we need
to provide for a conservation of the ionic currents that flow into
the axon as the pulse rises and out again as it passes by. One can
show that the equations

$$\frac{\partial v}{\partial t} + f(v) + \epsilon z = \frac{\partial^2 v}{\partial x^2}$$

$$\frac{\partial z}{\partial t} = v$$

$$f(v) = v(1 - v)(a - v)$$

represent such a system. The diffusion equation takes the place of the v and m equations of the Hodgkin-Huxley theory, and v now replaces both of those variables. The first-order equation in z replaces the first-order equations in h and n, and z stands for those variables. It may be shown that these simple equations result from approximating the basic experimental current-voltage measurements (called voltage clamp experiments [1, 16]), assuming that the sodium ion flow response is instantaneous [$\tau_m(v)$ in equation is taken equal to zero], and changing time and distance scales and redefining dependent variables. ϵ then is a ratio of times and has a value of the order of $0.01 - 0.001$, whereas a, which is a measure of the balance point at which ion inflow and outflow are equal and therefore is a threshold variable, is about $0.05 - 0.1$. z lies between zero and one and is always positive. $v(x, t)$ should have a pulse shape.

We have already seen that for $v = v(x + \theta t)$ and $\epsilon = 0$, there is an exact solution given in Table 1. There is another solution for $\epsilon = 0$. At $\theta = 0$,

$$(2.9) \quad v = \frac{2a\left[\left(\kappa e^{-\sqrt{a}\xi} + \frac{1+a}{3}\right) - \frac{1+a}{3}\right]}{\left(\kappa e^{-\sqrt{a}\xi} + \frac{1+a}{3} - \frac{\sqrt{2a}}{2}\right)\left(\kappa e^{-\sqrt{a}\xi} + \frac{1+a}{3} + \frac{\sqrt{2a}}{2}\right)},$$

$$\kappa = \frac{a}{v_0(0)} - \frac{1+a}{3}$$

satisfies the equation. It is a symmetrical function, and in Fig. 8 these solutions form a base line in the θ–a plane.

Now for $\epsilon > 0$, we would like $v(x, t) = v(\xi)$ to approach zero as $\xi \to \pm\infty$. This means that for a small change in ϵ, the boundary condition at $\xi \to +\infty$ must make a large change, from one to zero.

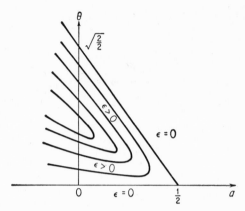

FIG. 8. Wave velocity as a function of the parameter a for the FitzHugh-Nagumo equation.

We may begin by continuing the solution for $\epsilon = 0$ by a regular expansion for $\epsilon > 0$, but this will not satisfy the singular condition at $\xi \to \infty$. What one must do is to retrieve, for $\epsilon > 0$, from the $\epsilon = 0$ solution as ξ approaches $+\infty$, all of the interesting aspects of the pulse return. To do this, we must stretch the ξ coordinate and look into the regions of order ξ/ϵ. In fact, this kind of singular perturbation analysis on an infinite interval shows that there are two regions of order ξ/ϵ and two of order one in ξ. The pulse, for

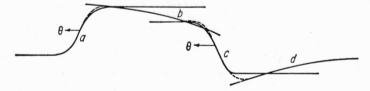

FIG. 9. The singular perturbation version of the FitzHugh-Nagumo traveling wave solution.

small ϵ, can be thought of as being composed of a front (a), a long (order ξ/ϵ) flat top (b), a back side (c), which must travel at the same velocity as the front but which has lower asymptotic values at infinity, and (d), a return to $+\infty$. The dotted curves indicate where asymptotic matching procedures must be used. Such a solution can be found. The value for θ will be altered only by an amount of order ϵ:

$$(2.10) \qquad \theta = \theta_0 + \frac{\epsilon}{\theta_0} \frac{\int_{-\infty}^{+\infty} e^{-\theta_0\xi'} \frac{dv_0}{d\xi'} \left(\int_{-\infty}^{\xi'} v_0(\xi'') \, d\xi'' \right) d\xi'}{\int_{-\infty}^{+\infty} e^{-\theta_0\xi'} \left(\frac{dv_0}{d\xi'} \right)^2 d\xi'},$$

where θ_0 and v_0 are the $\epsilon = 0$ values given in (f) of Table 1. For $a = 0$, $\theta_0 = \sqrt{2}/2$, which, in dimensional terms, comes out to be about 35 meters/sec. The actual velocity of the squid axon pulse for these parameters is 20 meters/sec. As a increases, the theoretical θ_0 decreases, but there is a limit as to how far θ_0 can go. One can see this by looking at numerical results for these equations. In Fig. 8, the curves inside the triangle show θ–a values for various ϵ. There is the strong suggestion that one of the two families of solutions for each value of a is stable (upper branch *is* stable for $\epsilon = 0$) and the other unstable (true for $\epsilon = 0$). Not only do the waveforms become increasingly nervepulselike in shape as ϵ increases, but James Cooley has shown numerically that these shapes are taken on asymptotically from the initial value problem for the original partial differential equations.

There is also a perturbation expansion for the lower branch solutions, but this is regular and, except near $a = \frac{1}{2}$, is an expansion in $\epsilon^{1/2}$.

Finally, if one looks at the singular perturbation solution for small ϵ again, one may ask if there are other solutions. Indeed, the forward moving front and equal velocity back side suggest that one may be able to fit other pulse amplitudes in this fashion. This is, in fact, the case. A set of periodic solutions can be found in which the period, the wave heights, and the velocity of propagation are all related. The details of this analytical solution are being carried out with Paco Lagerstrom at the California Institute of Technology.

One begins to see that the interestingly nonlinear phenomena that have come out of the empirically rich Hodgkin-Huxley theory may be usefully modeled by the simpler set of equations.

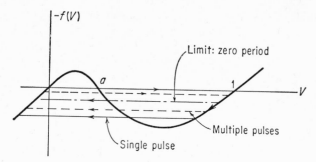

FIG. 10. Periodic solution to the FitzHugh-Nagumo equations can exist.

3. THE STEFAN PROBLEM

In a number of physical problems, concentration, temperature, magnetic field, or other field changes alter the state of substances in such a manner that phase transitions occur. The change of phase is usually characterized by the field variables reaching critical values and by a change in the energy state of the material. A typical example is the melting or freezing of a liquid, in which a melting or freezing temperature is reached and there is a latent heat of transition. The mathematical description of such problems involves parabolic equations, for temperature or concentration say, which may, indeed, be linear. However, the location at which the phase transition takes place is not known a priori. In general, this region or surface travels through the substance—there is a phase transition speed. This boundary location becomes a part of the solution, and since it is dependent on the solutions to the differential equations, the problem is nonlinear.

To be more explicit, consider a slab, which may be part ice and part water, infinite in two dimensions and finite in the third, say the x-direction. A controlled temperature at $x = 0$, the ice surface, causes the slab to freeze further. We shall take the equations for the temperature in the ice (region 1) and the water (region 2) to be

$$(3.1) \quad \frac{\partial u_1}{\partial t} = \frac{\partial^2 u_1}{\partial x^2}, \quad 0 < x < s(t), \qquad \frac{\partial u_2}{\partial t} = \alpha^2 \frac{\partial^2 u_2}{\partial x^2}, \quad s(t) < x < l.$$

$s(t)$ is the equation of the phase transition boundary, and we take

(3.2) $s(0) = s_0$ (the initial ice-water boundary), $0 \leq s_0 < l$.

For the other conditions, we have

(3.3) $u_1(0, t) = f_1(t) \leq 0,$ $u_2(l, t) = f_2(t) \geq 0,$

(3.4) $u_1(x, 0) = \phi_1(x) \leq 0,$ $0 < x \leq s_0$

(3.5) $u_2(x, 0) = \phi_2(x) \geq 0,$ $s_0 \leq x < l$

(3.6) $u_1(s(t), t) = 0 = u_2(s(t), t),$ $0 < t$

(3.7) $\phi_1(s_0) = \phi_2(s_0) = 0$

(3.8) $k_1 \dfrac{\partial u_1}{\partial x}\bigg|_{x = s(t)} - k_2 \dfrac{\partial u_2}{\partial x}\bigg|_{x = s(t)} = -L \dfrac{ds}{dt}.$

These equations are normalized so that, for example, α^2 is the ratio of thermal diffusivity in water to that in ice. k_1, k_2, L, and x, t also represent normalized quantities.

This is, then, a fairly general formulation that we shall refer to as the two-region problem. Notice that if $l \to \infty$, and $s_0 = 0$ and $u_2 \equiv 0$ ($\phi_2 \equiv 0$), the problem reduces to a one-region boundary value problem, and if $l \to \infty$, $s_0 = 0$, $u_1 \equiv 0$ ($\phi_1 \equiv 0$), the problem is a one-region initial value problem.

If we set

(3.9) $y = x - s(t),$ $t = t,$

then

(3.10) $\dfrac{\partial u_1}{\partial t} = \dfrac{\partial^2 u_1}{\partial y^2} + \dfrac{ds}{dt} \dfrac{\partial u_1}{\partial y},$

and the phase boundary conditions transform accordingly from conditions on $x = s(t)$ to conditions at $y = 0$. The term $\dfrac{ds}{dt} (\partial u_1 / \partial y)$ is effectively the convection of heat flux through the moving front. Since $\dfrac{ds}{dt}$ is unknown and is, in fact, determined by the functions u_1 and u_2, the nature of the nonlinearity can be seen.

For the case in which $l \to \infty$, $f_1 = u_0 < 0$ constant, $s_0 = 0$, and $\phi_2(x) = u_1 > 0$ constant, there is a similarity solution. With $\xi = xt^{-1/2}$ and $s = 2bt^{-1/2}$ one can easily solve the system of equations and obtain solutions for u_1 and u_2 with b determined by

$$(3.11) \quad k_1 u_0 e^{-b^2} \left(\int_0^{2b} e^{-\xi'^2/4} \, d\xi' \right)^{-1}$$
$$+ \, k_2 u_1 e^{-b^2/\alpha^2} \left(\int_{2b}^{\infty} e^{-\xi'^2/4\alpha^2} \, d\xi' \right)^{-1} = bL.$$

This problem is often referred to as the Stefan problem. It was actually first exhibited as a model of the cooling of the earth by Clapeyron and Lamé [17] and then used by Stefan [18] some years later. In the ensuing long period, the similarity solution has been applied in many areas (and has often been rediscovered in the course of these applications). There have been problems solved in missile nose-cone ablation, superconducting switching, soldering, crystal growth, applications to control theory, econometrics, geophysical boundary motion, and many others. A recent book by Rubinstein [19] summarizes many of the results of this theory.

The basic Stefan problem has had a number of interesting extensions and alterations worked on it in recent years. We shall describe some of them. Before doing so, we can also mention that there has been a qualitative treatment of certain classes of Stefan problems that deals with questions of existence, uniqueness, and formal asymptotic behavior. We shall not cite these results in any detail; they are fully dealt with in [19] and in [20]. It should be mentioned that for the most part existence and uniqueness have been shown only for small times, and little has been determined for the finite boundary, finite time of transition problem.

Although we shall deal with analytical approximations, the powerful tool in recent years for learning about nonlinear behavior in these problems has been a set of numerical solutions. The numerical methods will not be described, but a special group of references [21] is provided, and the results of numerical work are cited for comparison in some cases and explicit interest in others.

All of the examples are for the one-dimensional problem. Very little analytical work has been done on the two- or three-dimensional cases. There are some similarity solutions [22] and some approximate methods [23]. It is an open field, however, and some of the approximate methods given below should be applicable.

a. SMALL-TIME APPROXIMATIONS. In some cases the complete passage of a phase transition through a material is not required but rather only the initial behavior at the boundary. In other cases this initial behavior is useful as a starting solution away from a singularity or as a guide. Three methods have been used: (i) simple, direct expansion of the solution in a power series in x and t; (ii) more general expansion that allows for initial singularities; and (iii) integral equation formulations using heat equation Green's functions.

(i) The double series expansion for a single region problem is a straightforward application of this simple method of anlaytic expansion. For the single-region boundary problem defined by using only $u_1(x, t)$ in (3.1), (3.2), (3.3), (3.6), and (3.8) with $s_0 = 0$, we can obtain by expansion in a series in x and t ($L_1 = L/k_1$):

$$(3.12) \quad u_1(x, t) = f'(0)t - [L_1 f'(0)]^{1/2} x$$

$$+ f''(0) \frac{t^2}{2} - \{L_1 f''(0) + 2(f'(0))^2\} [L_1 f'(0)]^{-1/2} \frac{xt}{3}$$

$$+ f''(0) \frac{x^2}{2} + O(t^3), \qquad 0 < x < s(t)$$

$$(3.13) \quad s(t) = \left[\frac{f'(0)}{L_1}\right]^{1/2} t$$

$$+ \left\{ f''(0) - \frac{1}{L_1} (f'(0))^2 \right\} [L_1 f'(0)]^{-1/2} \frac{t^2}{6} + O(t^3).$$

If $f(t)$ is of the form $f(t) = mt + O(t^3)$, then

$$(3.14) \qquad s(t) = \left(\frac{m}{L_1}\right)^{1/2} t - \left(\frac{m}{L_1}\right)^{3/2} \frac{t^2}{6} + O(t_3) + \cdots.$$

Notice that $f(t)$ is taken to be continuous at $t = 0$; in fact, since $u_1(s(0), 0) = 0, f(0) = 0$ is required; no discontinuity is allowed at $x = t = 0$. This means that the classical similarity solutions that are singular at $t = 0$ are not described by these expansions. Another important observation is that initial value problems, in which $u_1(x, 0) = \phi(x)$ is prescribed for $x > 0$ instead of $f(t)$ along $x = 0$, do not allow such expansions. The characteristics for the heat equation are lines parallel to the x-axis, so attempts to extend a solution from $t = 0$ will conflict with the extensions from $x = s(t)$. This was overlooked in [24].

(ii) *More general expansions.* A remedy for both of the difficulties is to define a more general kind of expansion. In fact, we may try to combine the series and similarity methods.

Let us introduce new variables $\xi = xt^{-1/2}$ and η and take, because of the frequent dependence exhibited, $\eta = t^{1/2}$. Then the partial differential equation becomes

$$(3.15) \qquad \frac{\partial^2 u}{\partial \xi^2} + \frac{\xi}{2} \frac{\partial u}{\partial \xi} - \frac{\eta}{2} \frac{\partial u}{\partial \eta} = 0.$$

Because we expect ξ to play a prominent role for small times, we set

$$(3.16) \qquad u(\xi, \eta) = u_0(\xi) + \eta u_1(\xi) + \eta^2 u_2(\xi) + \cdots,$$

$$(3.17) \qquad s(\eta) = \eta \sigma(\eta) = \eta \sigma(0) + \eta \sigma_\eta(0) + \cdots.$$

To the lowest order, u_0 is the single-region similarity solution already referred to. To the next order the system can be solved to obtain

$$(3.18) \qquad u_1(\xi) = c_1 \xi + F_\eta(0)\left(e^{-\xi^2/4} + \xi \int_0^{\xi/2} e^{-w^2}\, dw\right)$$

with c_1 and $\sigma_\eta(0)$, determined by a set of equations similar to (3.11).

An application of this general idea has been attempted in connection with a problem that occurs in the geophysical studies of discontinuities that may arise from phase transitions inside the earth. The general problem is formulated in O'Connell and Wasserburg [25], and another kind of approximate solution as well as numerical solutions are provided there. The case we mention here, the "gutted" problem, has the peculiarity that the critical phase transition temperature depends on the location of the phase transition boundary; equation (3.6) becomes

$$u_1(s, t) = u_2(s, t) = Ds(t).$$

In this case, we keep $\xi = xt^{-1/2}$ but take $\eta = x$ for both $x > s(t)$ and $x < s(t)$. The differential equations become, for both u_1 and u_2 [if we let $u_{1,2}(x, t) \to u_{1,2}(\xi, \eta)$]:

$$(3.19) \qquad \frac{\partial^2 u}{\partial \xi^2} + 2 \frac{\eta}{\xi} \frac{\partial^2 u}{\partial \xi\, \partial \eta} + \left(\frac{\eta}{\xi}\right)^2 \frac{\partial^2 u}{\partial \eta^2} + \frac{\xi}{2} \frac{\partial u}{\partial \xi} = 0.$$

We take

(3.20) $u(\xi, \eta) = u_0(\xi) + \eta u_1(\xi) + \eta^2 u_2(\xi) + \cdots,$

where again u_0 will be the similarity solution. To the next order in η_1 for u_1, for example, we have

(3.21) $$\frac{\partial^2 u_1}{\partial \xi^2} + \frac{2}{\xi}\frac{\partial u_1}{\partial \xi} + \frac{\xi}{2}\frac{\partial u_1}{\partial \xi} = 0,$$

which has as a solution

(3.22) $$u_1(\xi) = K_1 \int \frac{1}{w^2} e^{-w^2/4}\, dw + K_2,$$

and the conditions given are sufficient to determine the constants of integration and the constants in the perturbation analysis.

The method thus serves to move a problem like the Connell-Wasserburg problem away from the simple similarity solution. It cannot carry such a problem into regions where the boundary can turn back, however—it cannot, for example, give the effects of a right-hand boundary. In the geophysical application, this is important. We shall point out some results that may be applicable to such problems a bit later.

(*iii*) *Integral equation formulations.* The general idea is to construct solutions to the heat equations of the Stefan problem by distributing an appropriate set of singularities along the plane $t = 0$. This is like distributing single and double layers in potential theory. The idea has been used in the qualitative theory [19] and [20] and may also be used to start small time expansion. It cannot, however, help much with the finite boundary transition problem.

Such a formulation does, however, suggest that an inverse method may be developed by choosing the unknowns and then calculating the resulting initial and boundary conditions. This has been suggested by Martynov [26, 27].

b. THE EFFECT OF THE FINITE BOUNDARY

(*i*) *Switching of a superconducting strip.* Since the small time expansions are not useful for gaining information near the far boundary, one might try to start an expansion in the neighborhood of this boundary. The difficulties are that there is a scale length l

in the problem that obviously is of importance near $x = l$. There need not be a similarity solution, and the "ending" solution's behavior is not known. However, there have been some numerical experiments that are interesting with respect to this question and also because they were done on an interesting application of the Stefan problem.

The application concerns the switching of a bulk superconducting material from the superconducting to the normal state. The state of superconducting materials depends on both temperature and magnetic field. If the temperature is held constant at a value below the critical temperature for switching when the magnetic field is zero, then an increase in the magnetic field will allow the material to change from the superconducting to the normal state. Suppose we consider a strip of superconductor at temperature T infinite in two dimensions but finite in the x-direction of length l. We shall apply, at $t = 0$, a constant magnetic field H_e at $x = 0$ and

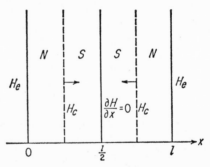

FIG. 11. The symmetric superconducting-to-normal phase transition.

$x = l$ with $H_e > H_c(T)$. This will cause a normal superconducting front to move into the strip. In the symmetrical case described, phase boundaries move from both $x = 0$ and $x = l$. If we choose as a length scale l and as a normalizing time $l^2 c^2 \pi \sigma^n$ (c is the velocity of light and σ^n is the ohmic conductivity of the normal region), then the London equations for the magnetic fields in the normal ($H^N(x, t)$) and superconducting ($H^S(x, t)$) regions are

$$(3.23) \qquad \frac{\partial^2 H^N}{\partial x^2} = \frac{\partial H^N}{\partial t}, \qquad\qquad 0 \leq x < s(t)$$

$$(3.24) \qquad \frac{\partial^2 H^S}{\partial x^2} = \alpha H^S + \beta \frac{\partial H^S}{\partial t}, \qquad s(t) < x \leq l.$$

For this case, the conditions along the moving and fixed boundaries are as follows:

At $x = s(t)$:

$$(3.25) \qquad\qquad H^N(s, t) = H^S(s, t) = H_c,$$

$$(3.26) \qquad \frac{\partial H^S}{\partial x}(s, t) - \frac{\partial H^N}{\partial x}(s, t) = -\alpha \int_{s(t)}^{l/2} H^S(\xi, t)\, d\xi.$$

At $x = l/2$:

$$(3.27) \qquad\qquad \frac{\partial H^S}{\partial x}\left(\frac{l}{2}, t\right) = 0.$$

At $x = 0$:

$$(3.28) \qquad\qquad H^N(0, t) = H_e.$$

At $t = 0$:

$$(3.29) \qquad H^S(x, 0) = \frac{H_0 \cosh \sqrt{\alpha}\left(\dfrac{l}{2} - x\right)}{\cosh \sqrt{\alpha}\,\dfrac{l}{2}}.$$

Werner Liniger [28] has computed difference equation solutions for this boundary value problem, and these are shown in Figs. 12 and 13. Notice that although the transition velocity is like $t^{-1/2}$ initially, it then slows down until it nears $x = l/2$, where it speeds up again. Notice also the difference between the finite region numerical solution and the numerical solution for a half space (Fig. 12).

(*ii*) *Melting of a lake.* Suppose we consider the surface of an ice-covered lake to be at $x = 1$ and the bottom of the lake at $x = 0$. Obviously, the melting as it really happens is a complex phenomenon—movements of the water, cracking of the ice, nonhomogeneities, etc., all playing a role—but let us consider a simple version within the scope of the Stefan problem. The results that follow have all been obtained numerically by using a method and a program prepared by Werner Liniger.

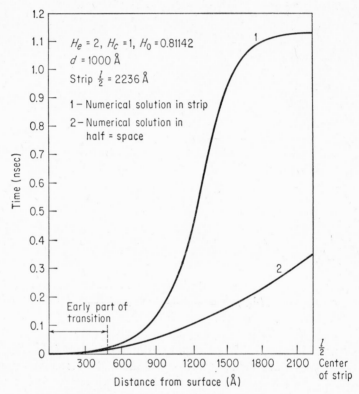

FIG. 12. Results of a numerical solution for the superconducting-to-normal transition. d is the intrinsic penetration distance of the material. The time is in nanoseconds.

Letting $u_2(x, t)$ and $u_1(x, t)$ represent the ice and water temperatures, respectively, we take as initial conditions

$$(3.19) \quad u_1(x, 0) = -\tfrac{16}{3} x + 4, \qquad u_2(x, 0) = -(x - \tfrac{3}{4}).$$

These are chosen so that $u_1(0.75, 0) = u_2(0.75, 0) = 0$; i.e., $s(0) = 0.75$. We shall assume that on the ice surface there is an inflow of heat given by

(3.20)
$$\frac{\partial u_2}{\partial x}(1, t) = \frac{t}{20} - 1,$$

and on the bottom, a constant temperature,

(3.21)
$$u_1(0, t) = 4.$$

All of the numerical values and the time scale have been normalized.

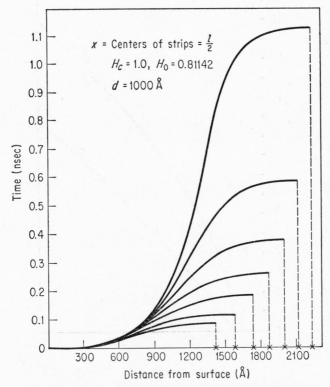

FIG. 13. Time of superconducting transition as a function of slab width.

For the conditions given, the ice will melt through, the phase boundary moving monotonically upward toward the surface. As one can see from Fig. 14 (curve a), the dimensionless time for melting is $t = 11$. Dimensionally this is $\tau = tx_0^2 \rho_1 c_1 / k_1$, where x_0 is the depth from ice surface to water bottom and ρ_1, c_1, k_1 are the density, specific heat, and thermal conductivity of water.

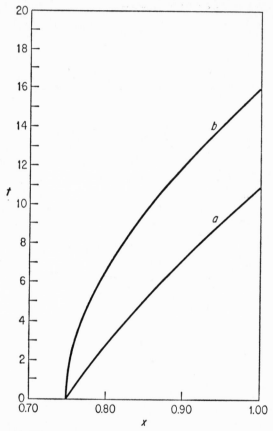

FIG. 14. Time of lake melting; a and b represent different boundary and initial conditions.

If the initial state is altered so that

$$u_1(x, 0) = -4x + 3, \qquad u_2(x, 0) = (x - \tfrac{3}{4}), \quad \text{and} \quad u_1(0, t) = 3,$$

then the ice melts again but more slowly [see Fig. 14 (curve b)]; $t = 16$.

In both of these cases all of the melting is from the bottom up; the ice surface reaches melting temperature only when the phase boundary arrives from below.

If instead we take (with exactly the same scaling)

$$u_1(x, 0) = u_2(x, 0) = 1 - 2x$$

so that $s_0 = 0.5$ and

$$u_1(0, t) = 1, \qquad \frac{\partial u_2}{\partial x}(1, t) = 2\left(\frac{t}{m} - 1\right),$$

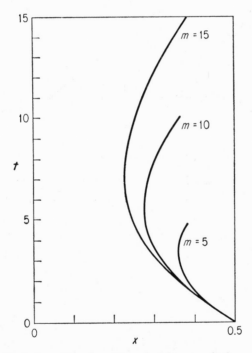

FIG. 15. Lake melting transitions that do not melt through.

then the melting curves look like those of Fig. 15. The ice-water interface actually moves down towards the bottom before it begins to move upward. The rate of melting is much slower, and one can see that the melting process will be very long. In fact, the top

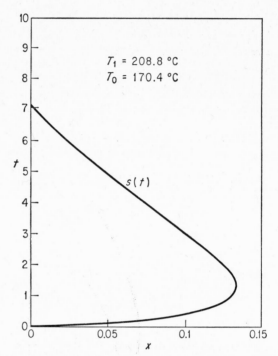

FIG. 16. Growth and decay in a dip soldering problem; T_1 and T_0 represent initial temperatures of the solid pin and the solder bath, respectively.

surface will reach the melting point, and melting will begin from the top down as well.

It turns out to be, in fact, rather interesting to manipulate this interface motion by altering the initial and boundary conditions. Its behavior, however, is hard to analyze, and the flexible numerical method for this two-region problem has been invaluable.

The "backward" motion of the boundary in the last case suggests that the Stefan model can describe certain growth and decay processes simultaneously. Such a case has been studied as a model for the process called dip soldering. Briefly speaking, if a cool pin of a transistor is immersed in a solder bath, the solder first freezes along its sides and then, as the bath heat is conducted through it, melts to form, finally, a thin, tinned layer. There will thus be an optimum time to pull out the pin. This growth and decay has been studied by Tadjbakhsh and Liniger [29] both numerically and by a singular perturbation method. If the frozen layer is small with respect to the bath size, one may expand solutions about a zero-thickness frozen region solution. The perturbation is singular because the equations that hold in the frozen region will not appear to the lowest order, although there will be boundary conditions remaining. In Fig. 16 one can see what the solutions look like, and in Fig. 17 a comparison is given of numerical and perturbation methods.

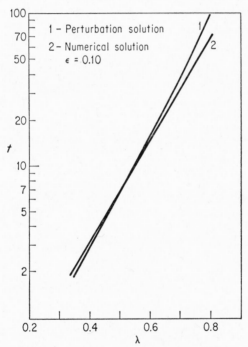

FIG. 17. Comparison between perturbation and numerical solutions for the dip soldering problem. ϵ is a measure of the heat energy available compared to the latent heat; λ is the ratio of heat energy in the frozen material to that of the bath.

(*iv*) *Periodic motion of the phase transition boundary.* Consider a two-region problem again for which at one boundary, $x = 0$, we impose a periodic temperature change $u_1(0, t) = f_1(t)$. Let the other side of the region be kept at a fixed temperature, $u_2(1, t) = -0.5$. Again, the thermal coefficients are taken to be those for water in region 1 and in region 2. In the cases that were computed numerically, the initial state of the region was taken to be ice at the critical temperature

$$u_2(x, 0) = 0$$

and

$$s(0) = 0.$$

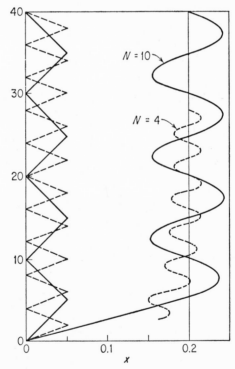

FIG. 18. Response of a phase transition boundary to periodic forcing. N is the period of the triangular forcing function.

The function $f_1(t)$ was a triangular function of the form shown in Fig. 18, and the maximum amplitude was kept fixed at one. The period N was varied from 0.1 to 10. Figures 18 and 19 show the calculated responses. The following observations can be made from these and other calculations that have been done:

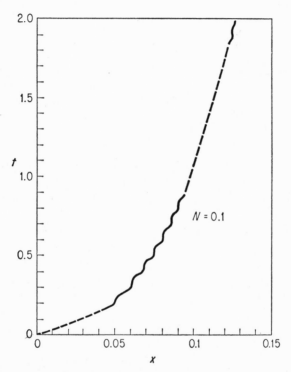

FIG. 19. Response of a phase transition boundary to a triangular forcing function of period $N = 0.1$.

1. After a transient period that becomes relatively longer as the period becomes shorter, the moving boundary obtains a periodic oscillation about a fixed position, the period being the same as that of $f_1(t)$. The fixed position depends on the mean amplitude of $f_1(t)$ and on the value of the fixed temperature at $x = 1$ in a simple, easily calculable manner.

2. For fixed amplitude of $f_1(t)$, the amplitude of the oscillation of $s(t)$ is a function of the frequency of $f_1(t)$. The phase lag of $s(t)$ also depends on this frequency.

These are linearlike responses. It is not clear whether there is resonancelike behavior; not enough numerical experiments have been done. When $f_1(t) = 0$ at the end of each period, since the phase boundary is also at $u_1(s, t) = 0$, the entire region between $x = 0$ and $x = s(t)$ must be at zero temperature.

When the periodic oscillation about the mean phase boundary line is small with respect to this location itself, the problem may be analyzed. In this case, the solution on each side may be set out separately, and then the heat balance equation may be used to identify coefficients in periodic series.

c. OTHER APPROACHES. There have been a number of other ideas for handling the Stefan problem. We shall simply mention several of these. One method, as we have mentioned earlier, is to solve the problem indirectly—to choose the boundary motions from which initial or boundary conditions can be calculated. Another approach is to use a moment method. If a complete set of functions $\phi(x, t)$ is multiplied into the partial differential equations and integrations performed, then integral relations or moment equations are formed. If there is a function that satisfies all of the integral relations formed by all ϕ of the complete set, then such a function is a solution of the equation. The Green and Neumann functions for the heat equation are examples of such weighting functions. If, as is practically the case, only a finite number of moment equations are used, an approximate solution is obtained. As an example, if we take again the one-region problem with $u_1(0, t) = f(t)$, multiply the differential equation by $\phi(x, t)$ and integrate by parts twice, we have

$$\frac{\partial}{\partial t} \int_0^s \phi u \, dx$$

$$= \int_0^s u \left(\frac{\partial^2 \phi}{\partial x^2} + \frac{\partial \phi}{\partial t} \right) dx - \phi(s, t)\kappa \dot{s} + \left(-\phi(0, t) + \frac{\partial \phi}{\partial x}(0, t) \right) f(t).$$

If $\phi = 1$,

$$\frac{\partial}{\partial t} \int_0^s u\, dx = -L\dot{s} - f(t).$$

As an example of the use of this equation, suppose we choose

$$u = h(t)\left(1 - \frac{x}{s}\right).$$

Then the differential equation yields

$$s(t) = \frac{1}{\left(L + \dfrac{h}{2}\right)} \left[s_0^2 \left(L + \frac{h(0)}{2}\right)^2 + 2 \int_0^t \left(L + \frac{h}{2}\right) h\, dt \right]^{1/2}.$$

If, in particular, $s_0 = 0$ and $h = $ constant,

$$s(t) = 2\left(1 + \frac{2L}{h}\right)^{-1/2} t^{1/2}.$$

The exact solution for this problem is

$$s(t) = p t^{1/2},$$

where p satisfies an equation like (3.11). When $L/h \gg 1$, $p = (2h/L)^{1/2} + 0(h/L)$, which agrees with the approximation.

If $h(t) = 1 - \cos \omega t$ and $s_0 = 0$, then

$s(t)$

$$= \frac{\sqrt{2}}{L + \dfrac{1 - \cos \omega t}{2}} \left[t\left(L + \frac{3}{4}\right) - \frac{\sin \omega t}{\omega}(1 + L) + \frac{1}{8\omega} \sin 2\omega t \right]^{1/2}.$$

For small t, this yields

$$s(t) = \frac{1}{\sqrt{3L}} \omega t^{3/2}.$$

For large t:

$$s(t) = \frac{\sqrt{2t\left(L + \frac{3}{4}\right)}}{L + \dfrac{1 - \cos \omega t}{2}} \left[1 - \frac{\dfrac{\sin \omega t}{\omega}(1 + L) - \dfrac{1}{8\omega}\sin 2\omega t}{2t\left(L + \dfrac{3}{4}\right)} \right].$$

This is not quite the kind of solution exhibited for the numerical case described earlier. Here the boundary oscillates about $s \sim \sqrt{t}$, and the amplitude of oscillation continues to increase with time. However, it is approximately correct for values of t not too large.

The moment methods can clearly be carried out for many cases. They will not be further discussed here.

There are still other general notions such as those of Boley [30] on imbedding the unknown boundary problem into a larger boundary value problem. Upper and lower bounds on solutions can be obtained in this fashion [31]. There is also a group of problems in which the latent heat does not appear in the free boundary condition. These require new approaches and have been considered by Sherman [32].

4. CONCLUSION

We have tried, by giving some very specific examples, to point out how nonlinearity affects diffusion problems. Although the methods mentioned are not especially complex, many of them allow for both quantitative and qualitative extensions. The purpose of these kinds of analytical inquiries ought to be carefully understood. As we have shown, numerical methods for digital calculation are both powerful enough and flexible enough to be used experimentally. The value of approximate analysis cannot be overplayed, therefore. It can be used to provide starting solutions or to help to organize quantitative results or, most importantly, to grasp more fully the relationship between the physical aspects of the model and the mathematics. In the problems discussed here, the nonlinear effects are exhibited in the appearance of steady self-propagating pulses and of transition fronts.

In collecting the examples discussed and in doing the analysis of some of them, the author has had the pleasure of collaborating with several people. On the nerve-pulse problems these have included James Cooley and Fred Dodge of IBM Research and on the Stefan problem, Joseph Keller of the Courant Institute of New York University and Werner Liniger of IBM Research.

REFERENCES

1. Hodgkin, A. L. and A. F. Huxley, *J. Physiol.*, **117** (1952), 500–544.

2. Spalding, D., *Phil. Trans.*, **1** (1956), A 249.

3. Zeldovich, Y. and G. Barenblatt, *Combustion and Flame*, **3** (March 1, 1959), 61–74.

4. Fisher, R., *Annals of Eugenics*, **7**, Pt. 4 (1937), 355.

5. Johnson, W., *Arch. for Rat. Mech. and Anal.*, **13**, No. 1 (1963), 46–54.

6. Kolmogoroff, A., I. Petrovsky, and N. Piskunov, *Moscow Univ. Bull. Math.*, *Sèrie Internationale*, Sec. A, Math. et Mèc., **1**, No. 6 (1937), 1.

7. Kanely, I., *Math. Sbornik*, **59** (Supplement, 1962), 245–288.

8. FitzHugh, R., *J. of Gen. Physiol.*, **43** (1960), 867–895.

9. Noble, D., *Physiol. Reviews*, **46**, No. 1 (1966), 1–50.

10. Huxley, A., *J. of Physiol.*, **148** (1959), 80–81.

11. Cooley, J. and F. Dodge, *Biophys. J.*, **6**, No. 5 (1966), 583–599.

12. Sabah, N. and K. Leibovic, *Biophys. J.*, **9**, No. 10 (1969), 1206–1222.

13. Mauro, A., F. Conti, F. Dodge, and R. Schorr, to be published in *J. Gen. Physiol.* in press.

14. FitzHugh, R., *Biophys. J.*, **1** (1961), 445–466.

15. Nagumo, J., S. Arimoto, and S. Yoshizawa, *Proc. IRE* (1962), 2061–2070.

16. Cole, K. S., *Ions, Membrane and Impulses.* Berkeley, Calif.: University of California Press, 1968.

17. Clapeyron, B. P. E. and G. Lamé, *Annales de Chimie et de Physique t.*, **XLVII** (1831), 250–256.

18. Stefan, J., *Sitzungsberichte der Akad. Wiss. Wien. Math. Naturwiss.-Geb.*, **98** (Abth II) (1889), 473–484.

19. Rubinstein, L., *The Stefan Problem (Problema Stefan).* Latvian National Univ. in the name of Petra Stuchki, Computing Centre: Publisher Zvaigshne Rign., 1967 (in Russian).

20. Friedman, Avner, *Partial Differential Equations of Parabolic Type.* Englewood Cliffs, N.J.: Prentice-Hall, Inc., 1964.

21. Crank, J., *Quart. Jour. Mech. and Applied Math.*, **X** (1957), 220–231.
    de G. Allen, D. N. and R. T. Severn, *Quart. Jour. Mech. and Applied Math.*, **XV** (1962), 53–62.
    Douglas, J., Jr., and T. M. Gallie, Jr., *Duke Math. J.*, **22** (1955), 557–571.
    Ehrlich, L. W., *J. Assoc. Comp. Mach.*, **5**, No. 2 (1958), 161–176.
    Springer, G. S. and D. R. Olson, presentation at Winter Annual Meeting, New York, Nov. 25–30, 1962, of the American Soc. of Mech. Eng., Paper No. 62-WA-246.

22. Ham, F. S., *Quart. Appl. Math.*, **XVII** (1959), 137.

23. Poots, G., *Int. J. Heat Mass Transfer*, **5** (1962), 339–348.

24. Evans, G. W. II, E. Isaacson, and J. K. L. MacDonald, *Quarterly of Appl. Math.*, **VIII** (1950), 3.

25. O'Connell, R. and G. Wasserburg, *Rev. of Geophysics*, **54** (1967), 329–410.

26. Martynov, G. A., *J. of Technical Physics*, **XXV** (1955), 1754–1767.

27. Martynov, G. A., translated from *Zhurnal Tekhnicheskoi Fiziki*, **2**, No. 2 (Feb. 1960), 239–241.

28. Liniger, W., *J. of Math. Phys.*, **3**, No. 3 (May/June 1962), 578–586. See also H. Cohen, *Proc. Eighth Int. Conf. on Low Temp. Physics*, London, Sept. 16–22, 1962 (Butterworths, 1963), pp. 372–375.

29. Tadjbakhsh, I. and W. Liniger, *Quart. J. Mech. and Appl. Math.*, **XVII**, No. 2 (1964), 141–155.

30. Boley, B. A., *J. of Math. and Phys.*, **XL**, No. 4 (Dec. 1961), 300–313.

31. Boley, B. A., *Quarterly of Appl. Math.*, **XXI**, No. 1 (April 1963), 1–11.

32. Sherman, B., Research Report No. 68-7, May 1968, Rocketdyne North American Aviation, Inc., Canoga Park, Calif.

COMPUTER POWER AND ITS IMPACT ON APPLIED MATHEMATICS

Donald Greenspan

1. INTRODUCTION

The development of the high-speed digital computer has had and continues to have a revolutionary effect on modern applied mathematics. Immediate evidence is available in the form of a large number of computer-generated numerical solutions of fundamental, unsolved systems of mathematical equations. The diversity of fields being affected includes lunar and planetary astrodynamics [14, 39, 52], wave diffraction [2, 22, 27, 49], shock waves [22, 35, 42, 43], laminar flow of liquids [8, 24, 25e, 31, 45], free surface fluid flow [3], weather prediction [23, 36, 46], thermo-dynamics [2, 18, 22, 23, 29, 30, 49], elasticity [10, 23, 49, 51], electrostatic and gravitational potential [2, 16, 22, 23, 25c, 51], optimal control [19], n-body problems [21, 25g, 40], vibration theory [7, 25f, 49], molecular interaction [1, 21], quantum theory [23], and relativistic collapse [37].

In this paper, we shall attempt to illustrate some of the above applications and to develop some of the dramatic qualitative effects computers are having on applied mathematics. This will be done by exploring typical examples from three general areas of computer activity, namely, the discrete approach to physical problems, the approximate solution of unsolved continuous linear problems, and the approximate solution of unsolved continuous nonlinear problems.

2. DISCRETE MODEL THEORY

Discrete model theory is an approach to mathematical simulation that emphasizes *finite* structure and *constructability*. In it, one arithmetizes physical concepts and laws and uses the arithmetic power of the computer to solve both linear and nonlinear problems with great speed and ease. Existence and uniqueness theorems are almost immediate consequences of the formulation and require no tools from topology or functional analysis for their proof.

Let us illustrate the above remarks by outlining a discrete model approach to plane Newtonian dynamics.

Mathematically, the fundamental concepts of Newtonian mechanics, like *particle, time,* and *motion,* can be left undefined. Nevertheless, for the sake of intuition, let us take, typically, the concept of *motion* and develop, heuristically, a convenient and natural way of thinking about it. Our approach will be the physiological one which is associated with the viewing of motion pictures. In this situation a finite sequence of stills projected with rapidity on a screen and transferred as retinal images to the brain result in observed motion. To abstract this quantized approach to motion, we proceed as follows. Let Δt be a positive constant, which, in the above context, would represent the time interval between successive stills. For $k = 0, 1, 2, \cdots, n$, where n is a positive integer, let $t_k = k\, \Delta t$. The numbers t_0 and t_n may be considered to be initial and terminal times, respectively, of some physical event. Suppose that on an X-axis, a particle P is located at x_k at time t_k, $k = 0, 1, \cdots, n$. Then the motion of P from x_0 to x_n is merely P's being at $x_0, x_1, \cdots, x_n$ at the respective times t_0, t_1, $\cdots$, t_n. Thus, the motion of P from x_0 to x_n is considered to be a sequence of stills.

Next, in order to define the concepts of velocity and acceleration for a particle in motion, let us, for simplicity, confine our attention to motion in a fixed direction. Let P be constrained to move on an X-axis and let it be located at x_k at time t_k, $k = 0$, 1, $\cdots$, n. We wish to define P's velocity $v_k = v(t_k)$ and acceleration $a_k = a(t_k)$, for $k = 1, 2, \cdots, n$. For this purpose, consider first v_1 and a_1. Suppose one knows v_0 in addition to t_0, t_1, x_0, and x_1. For example, when a particle's motion begins from a rest position, one would know that $v_0 = 0$. In defining v_1, we shall try to use *all* the known data. If one were then to elect the elementary backward difference formula

$$(2.1) \qquad v_1 = \frac{x_1 - x_0}{\Delta t},$$

then, indeed, v_0 would have been neglected. If one were to elect the elementary forward difference formula

$$(2.2) \qquad v_1 = \frac{x_2 - x_1}{\Delta t},$$

then v_1 could not be calculated until one knew x_2. This is disturbing insofar as x_2 is not known at time t_1. But, also, the use of (2.2) would imply

$$(2.3) \qquad v_0 = \frac{x_1 - x_0}{\Delta t},$$

which is equally undesirable, for $v_0 = 0$ in (2.3) yields $x_0 = x_1$. Thus, for example, for such rudimentary problems like those of falling bodies, one would have the physically unacceptable implication that a body falling from a position of rest does not move when t changes from t_0 to t_1.

The method by which we shall utilize all the given data and yet avoid the shortcomings of (2.1) through (2.3) is to define v_1 implicitly by the smoothing formula

$$(2.4) \qquad \frac{v_0 + v_1}{2} = \frac{x_1 - x_0}{\Delta t}.$$

With regard to the acceleration a_1, under the above assumption that only x_0, x_1, t_0, t_1, and v_0 are known, it is reasonable to assume that

$$(2.5) \qquad\qquad a_1 = \frac{v_1 - v_0}{\Delta t}.$$

Motivated by (2.4) and (2.5), v_k and a_k will be defined in general by

$$(2.6) \qquad \frac{v_{k-1} + v_k}{2} = \frac{x_k - x_{k-1}}{\Delta t}, \qquad k = 1, 2, \cdots, n$$

$$(2.7) \qquad\qquad a_k = \frac{v_k - v_{k-1}}{\Delta t}, \qquad k = 1, 2, \cdots, n.$$

As will be seen later, these seemingly unrelated physically motivated formulas will combine in a most fortuitous way to yield the usual energy relationships of Newtonian mechanics. For the present, however, note that (2.6) and (2.7) can be written equivalently [25h] as

$$(2.8) \quad v_1 = \frac{2}{\Delta t} [x_1 - x_0] - v_0,$$

$$(2.9) \quad v_k = \frac{2}{\Delta t} [x_k + (-1)^k x_0 + 2 \sum_{j=1}^{k-1} (-1)^j x_{k-j}] + (-1)^k v_0,$$
$$k \geq 2,$$

$$(2.10) \quad a_1 = \frac{2}{(\Delta t)^2} [x_1 - x_0 - v_0 \Delta t],$$

$$(2.11) \quad a_2 = \frac{2}{(\Delta t)^2} [x_2 - 3x_1 + 2x_0 + v_0 \Delta t]$$

$$(2.12) \quad a_k = \frac{2}{(\Delta t)^2} \{x_k - 3x_{k-1} + 2(-1)^k x_0 + 4 \sum_{j=2}^{k-1} [(-1)^i x_{k-j}]$$
$$+ (-1)^k v_0 \Delta t\}, \qquad k \geq 3.$$

It is necessary now to have a discrete analogue of Newton's dynamical equation of motion. For this purpose, note that the right-hand side of each of (2.10) through (2.12) is a *linear* combination of x_i's. Thus, (2.10) can be solved easily for x_1, (2.11) can be solved easily for x_2, and (2.12) can be solved easily for x_k, $k \geq 3$. For this reason, we choose the dynamical difference equation

$$(2.13) \qquad ma_k = F(x_{k-1}, v_{k-1}, t_{k-1}), \qquad k \geq 1,$$

for then one can solve (2.13) for x_k easily and uniquely in terms of $x_{k-1}, x_{k-2}, \cdots, x_0, v_0, t_{k-1}$. By this means, (2.13) becomes a recursion formula for x_k, $k \geq 1$. Thus, given x_0 and v_0, one can use the arithmetic power of the computer to generate the sequence x_k, $k \geq 1$, whether F is linear or nonlinear. This sequence is called the solution of (2.13). Hence, our initial motivation for the choice (2.13) is computational.

Before discussing further physical implications of our discrete dynamical development, let us give an example of a prototype dynamical problem of nonlinear mechanics. As shown in Fig. 1,

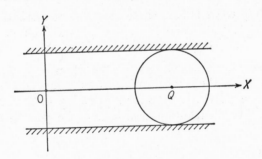

FIG. 1.

a particle of unit mass is constrained to move with its center Q on the X-axis. A displacement of the particle, such that the directed distance OQ is x, is opposed by a field force of magnitude $\sin x$ and a viscous damping force of magnitude αv, $\alpha > 0$. Then, from (2.13), the equation of motion of Q is

$$(2.14) \qquad a_k + \alpha v_{k-1} + \sin x_{k-1} = 0, \qquad k \geq 1.$$

Substitution of (2.8) through (2.12) into (2.14) yields

$$(2.15) \quad x_1 = x_0 + v_0 \, \Delta t - \frac{(\Delta t)^2}{2} [\alpha v_0 + \sin x_0],$$

$$(2.16) \quad x_2 = 3x_1 - 2x_0 - v_0\,\Delta t - \frac{(\Delta t)^2}{2}\left\{\alpha\left[\frac{2}{\Delta t}(x_1 - x_0) - v_0\right]\right.$$
$$\left. + \sin x_1\right\},$$

$$(2.17) \quad x_k = (3 - \alpha\,\Delta t)x_{k-1} + (-1)^{k-1}(2 - \alpha\,\Delta t)x_0$$
$$+ (2\alpha\,\Delta t - 4)\sum_{j=2}^{k-1}\left[(-1)^j x_{k-j}\right]$$
$$+ (-1)^{k-1}\left(1 - \frac{\alpha\,\Delta t}{2}\right)v_0\,\Delta t - \frac{(\Delta t)^2}{2}\sin x_{k-1}; \qquad k \geq 3.$$

Let us now prove that the solution of an initial value problem for (2.14) always exists and is unique.

THEOREM 2.1: If x_0 and v_0 are given, the solution of (2.15) through (2.17) exists and is unique for each positive α and Δt.

Proof: The proof is immediate from the recursive structure of (2.15) through (2.17) and from the observation that the right-hand sides of these equations are always defined and real.

Suppose next we actually set $x_0 = \pi/4$, $v_0 = 0$, $\alpha = 0.3$, and $\Delta t = 0.01$. Then the solution of (2.15) through (2.17) was generated on the UNIVAC 1108 for $n = 15{,}000$ in 33 seconds. The output was graphed automatically, and the portion between t_0

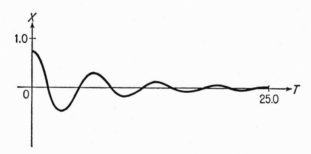

FIG. 2.

and t_{2500} is shown in Fig. 2. The motion exhibits strong damping and peak values of 0.754615, -0.480136, 0.297837, -0.185740, 0.116061, -0.072585, 0.045407, and -0.028421, which occur at times t_0, t_{327}, t_{648}, t_{966}, t_{1284}, t_{1602}, t_{1919}, and t_{2236}, respectively, and the time required for the particle to travel from one peak to the next decreases monotonically. Further details of the results are available in [25h], and the condition for stability is given in [13]. It is worth noting, in addition, that the above example is a discrete model formulation of the classical nonlinear pendulum problem [25h].

Returning now to the theoretical aspects of our discrete mechanics, let us show how the usual energy relationships can be developed. In the notation already developed, let F_i be the force applied at x_i, $i = 0, 1, \cdots, n$, in moving a particle P from x_0 to x_n. Then the work W done by F in moving P from x_0 to x_n is defined to be

$$W = \sum_{i=1}^{n} [(x_i - x_{i-1})F_{i-1}].$$

Then, from (2.6), (2.7), and (2.13),

$$W = m \sum_{i=1}^{n} (x_i - x_{i-1})a_i$$

$$= m \sum_{i=1}^{n} \left[(x_i - x_{i-1})\left(\frac{v_i - v_{i-1}}{\Delta t}\right) \right]$$

$$= m \sum_{i=1}^{n} \left[\left(\frac{x_i - x_{i-1}}{\Delta t}\right)(v_i - v_{i-1}) \right]$$

$$= m \sum_{i=1}^{n} \left[\left(\frac{v_i + v_{i-1}}{2}\right)(v_i - v_{i-1}) \right]$$

$$= \frac{m}{2} \sum_{i=1}^{n} [v_i^2 - v_{i-1}^2]$$

$$= \tfrac{1}{2}mv_n^2 - \tfrac{1}{2}mv_0^2.$$

The quantity

$$(2.18) \qquad\qquad K_i = \tfrac{1}{2}mv_i^2$$

is defined to be the kinetic energy of the particle at time t_i. The basic formula (2.18) results because the choices (2.6), (2.7), and (2.13) yield the telescopic sum

$$\frac{m}{2} \sum_{i=1}^{n} [v_i^2 - v_{i-1}^2]$$

in the above derivation of the formula for W. Since the kinetic energy formula (2.18) is now available, all the well-known conservation laws then follow in the usual way [25h].

One can now easily extend the discrete mechanics formulation to rotational problems [25h], to problems involving many particles [25f], and to problems in more than one dimension [25g]. Figure 3 shows the discrete dynamical motion of a nonlinear,

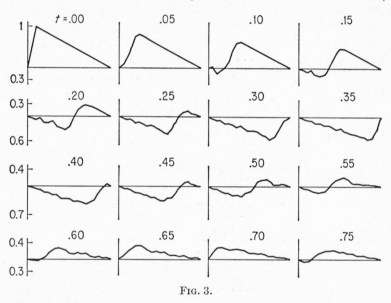

Fig. 3.

20 particle vibrating string during its first 0.75 sec of motion after release from its initial position of tension [25f]. For examples of discrete dynamical motion for nondegenerate three-body problems, see [25g].

For other discrete model studies, the reader should examine [5], [17], [20], [21], [33], [38], [42], [48], [50], and [58].

3. THE CAPACITY PROBLEM

Discrete model theory is at present in its infancy, and most contemporary applied mathematicians are concerned with traditional continuous models and techniques. Indeed, the greater portion of contemporary numerical analysis is devoted to approximating solutions of continuous problems. In this spirit, then, let us show how to approximate the solution of an unsolved continuous linear problem by means of a digital computer, and since initial value problems dominated the discussion in Sec. 2, we shall, for variety, turn our attention to a classical boundary value problem, the capacity problem [47], which can be formulated in electrostatic terms as follows.

Let S be a three-dimensional surface on which one wishes to obtain a given potential, which, for convenience, is taken to be unity. The problem is to determine how much charge C, called the capacity, one must put on S to obtain the given potential.

Analytically, the capacity problem can be formulated as follows. Let S be a three-dimensional surface that is closed, is bounded, contains the origin of XYZ space in its interior, and is homeomorphic to the unit sphere. Let R^* be the exterior of S and let $f(x, y, z) \equiv 1$ on S. Then if $u(x, y, z)$ is the solution of the exterior Dirichlet problem for the Laplace equation with boundary function f, one must (see [25a], [47]) evaluate:

$$(3.1) \quad C = -\frac{1}{4\pi} \iint\limits_{S} \frac{\partial u}{\partial n} \, dA$$
$$= \lim_{(x^2+y^2+z^2) \to \infty} [(x^2 + y^2 + z^2)^{1/2} \, u(x, y, z)].$$

Unfortunately, for a given nonspherical S, the precise numerical value of C is, in general, so difficult to determine that even the capacity of the unit cube has become a quantity of great interest [25a], [47]. Mathematicians have approached such problems by means of isoperimetric inequalities, whereas physicists and engineers have been prone to apply infinite series techniques [47]. The isoperimetric inequality approach requires special results for each S and yields only upper and lower bounds for C that are rarely sharp. The infinite series approach usually requires very extensive tables for each S, and these may have to be so voluminous in order to obtain a reasonable accuracy that the method often becomes impractical.

In formulating a numerical approach, then, to the study of capacity, note first that fundamental in the analytical formulation is an exterior Dirichlet problem. Let us then begin our development in the simplest possible way by summarizing the currently popular finite difference numerical method for Dirichlet problems. For simplicity we shall begin with two-dimensional problems.

Let G be a plane, bounded point set whose interior R is simply connected and whose boundary S is piecewise regular. If $f(x, y)$ is defined and continuous on S, the *interior* Dirichlet problem, called for simplicity the Dirichlet problem, for the Laplace equation

$$(3.2) \qquad \frac{\partial^2 u}{\partial x^2} + \frac{\partial^2 u}{\partial y^2} = 0$$

is that of determining a function $u(x, y)$ which is (a) defined and continuous on $R \cup S$, (b) is a solution of (3.1) on R, and (c) is identical with $f(x, y)$ on S. It is known that the solution of the Dirichlet problem exists and is unique. But because the solution is rarely constructable, the following numerical method has proved to be of exceptional practical value.

First let us approximate R by a finite subset R_h of R, and let us approximate S by a finite subset S_h of S. Intuitively, we shall define R_h to be a finite set of points like those shown to be crossed in Fig. 4, and we shall define S_h to be a finite set of points like

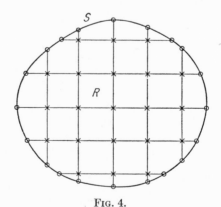

Fig. 4.

those shown to be circled in Fig. 4. More exactly, let $(\overline{x}, \overline{y})$ be an arbitrary, but fixed point in the plane, and let h be a positive constant called the grid size. The set of points $(\overline{x} + ph, \overline{y} + qh)$, $p = 0, \pm 1, \pm 2, \cdots, q = 0, \pm 1, \pm 2, \cdots$ is called a set of planar grid points. The set of vertical lines $x = \overline{x} + ph$, $p = 0, \pm 1, \pm 2, \cdots$, and of horizontal lines $y = \overline{y} + qh$, $q = 0, \pm 1, \pm 2, \cdots$, is called a planar lattice. Those planar grid points that are points of R are called interior lattice (or grid) points and are denoted by R_h. Let the set of points that S and the planar lattice have in common be denoted by S_h^* and set $G^* = R_h \cup S_h^*$. Let the four neighbors of a point (x, y) in R_h be defined as those four points in G_h^* that are closest to (x, y) in the east, north, west, and south directions. Let G_h be that subset of G_h^* which consists of each point of R_h and its four neighbors. Then, finally, the boundary lattice (or grid) points, denoted by S_h, are defined by $S_h = G_h - R_h$.

Next, at each interior point (x, y) of R_h, write down the Laplace difference equation approximation

$$(3.3) \quad -\left(\frac{2}{h_1 h_2} + \frac{2}{h_3 h_4}\right) u(x, y) + \frac{2}{h_1(h_1 + h_2)} u(x + h_1, y)$$

$$+ \frac{2}{h_2(h_1 + h_2)} u(x - h_2, y) + \frac{2}{h_3(h_3 + h_4)} u(x, y + h_3)$$

$$+ \frac{2}{h_4(h_3 + h_4)} u(x, y - h_4) = 0,$$

where, as shown in Fig. 5, $(x + h_1, y)$, $(x, y + h_3)$, $(x - h_2, y)$, $(x, y - h_4)$ are the neighbors of (x, y). If a neighbor of (x, y) is a point of S_h, then in (3.3) one substitutes the known boundary value $u \equiv f$. There results then a linear algebraic equation for each point of R_h. This system is solved by iteration, and the solution represents the numerical solution of the given analytical problem.

The numerical method outlined above has a firm mathematical basis in that the algebraic system always has a unique solution, a convergent iterative method can always be found to solve the system, and, as h goes to zero, convergence of the numerical to the analytical solution can be proved for a large class of problems

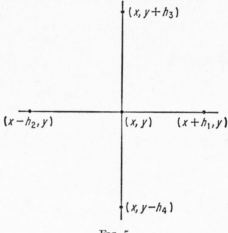

FIG. 5.

[25c]. That the method is computationally feasible follows from the observation that a resulting linear algebraic system of 25,000 equations can be solved on the UNIVAC 1108 in approximately 13 minutes. Finally, it is important to observe that the method extends in a natural way to any number of dimensions and to any elliptic equation for which the maximum principle is available [25c]. For three dimensions, in particular, the point sets R_h and S_h are defined as above, except that each point of R_h now has six neighbors and one applies the three-dimensional Laplace difference approximation

$$-\left(\frac{2}{h_1 h_2} + \frac{2}{h_3 h_4} + \frac{2}{h_5 h_6}\right) u(x, y, z) + \frac{2}{h_1(h_1 + h_2)} u(x + h_1, y, z)$$

$$+ \frac{2}{h_2(h_1 + h_2)} u(x - h_2, y, z) + \frac{2}{h_3(h_3 + h_4)} u(x, y + h_3, z)$$

$$+ \frac{2}{h_4(h_3 + h_4)} u(x, y - h_4, z) + \frac{2}{h_5(h_5 + h_6)} u(x, y, z + h_5)$$

$$+ \frac{2}{h_6(h_5 + h_6)} u(x, y, z - h_6) = 0.$$

Let us show now how the numerical method described for the

Dirichlet problem can be applied in a completely general and efficient way to estimate the capacity of any given surface S. Under the inversion mapping

$$(3.4) \quad x = \frac{\xi}{\xi^2 + \eta^2 + \nu^2}, \qquad y = \frac{\eta}{\xi^2 + \eta^2 + \nu^2},$$

$$z = \frac{\nu}{\xi^2 + \eta^2 + \nu^2}; \qquad \xi^2 + \eta^2 + \nu^2 \neq 0,$$

let us first transform the exterior problem, by means of which capacity is defined, into an interior problem. Let

$$R^* \to R^i,$$

$$S \to S^i,$$

$$(3.5) \qquad v(\xi, \eta, \nu) = (x^2 + y^2 + z^2)^{1/2} u(x, y, z),$$

$$F(\xi, \eta, \nu) = (\xi^2 + \eta^2 + \nu^2)^{-1/2}.$$

Then it is known [25c] that $v(\xi, \eta, \nu)$ is the solution of the *interior* Dirichlet problem on $R^i \cup S^i$ with boundary function $F(\xi, \eta, \nu)$. Moreover, it follows readily from (3.1), (3.4), and (3.5) that $C = v(0, 0, 0)$, and it is this important result that makes the approximation of C very simple on a computer. Indeed, in the case of a unit cube, for example, seven minutes of running time on the UNIVAC 1108 with a grid size of 0.045 yielded $C = 0.661$. For the same surface, approximately 40 years of work with isoperimetric inequalities (see [25a] for complete references) have yielded the bounds $0.654 < C < 0.6626$. For the approximation of the capacity of lenses, toroids, and other surfaces of physical interest, see [24a].

4. CAVITY FLOW

Finally, let us examine the impact of the modern digital computer on what is considered by many to be the most significant and difficult area of applied mathematics, that of continuous *nonlinear* models. Here, existence theorems, uniqueness theorems, and analytical methods for solving problems are rarely available.

For variety, let us examine a particular class of fluid problems, called cavity flow problems, which are of wide interest and can be formulated as follows. Let the points $(0, 0)$, $(1, 0)$, $(1, 1)$ and $(0, 1)$ be denoted by A, B, C, and D, respectively, as shown in Fig. 6.

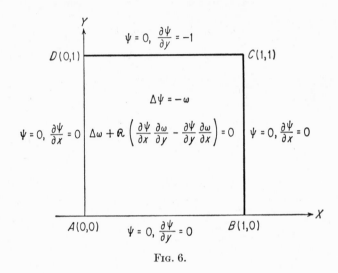

Fɪɢ. 6.

Let S be the square whose vertices are A, B, C, and D, and denote its interior by R. On R the equations of motion to be satisfied are the coupled, two-dimensional, steady-state Navier-Stokes equations; that is,

$$(4.1) \qquad\qquad \Delta\psi = -\omega$$

$$(4.2) \qquad \Delta\omega + \Re\left(\frac{\partial\psi}{\partial x}\frac{\partial\omega}{\partial y} - \frac{\partial\psi}{\partial y}\frac{\partial\omega}{\partial x}\right) = 0,$$

where ψ is the stream function, ω is the vorticity, and $\Re$ is the Reynolds number. On S the boundary conditions to be satisfied are

$$(4.3) \qquad\qquad \psi = 0, \qquad \frac{\partial\psi}{\partial x} = 0, \text{ on } AD;$$

$$(4.4) \qquad\qquad \psi = 0, \qquad \frac{\partial\psi}{\partial y} = 0, \text{ on } AB;$$

$$(4.5) \qquad \psi = 0, \qquad \frac{\partial \psi}{\partial x} = 0, \text{ on } BC;$$

$$(4.6) \qquad \psi = 0, \qquad \frac{\partial \psi}{\partial y} = -1, \text{ on } CD.$$

For given $\mathcal{R} \geq 0$, the mathematical problem is to solve (4.1) and (4.2) for ψ and ω subject to boundary conditions (4.3) through (4.6). For a restricted range of $\mathcal{R}$, the mathematical problem (4.1) through (4.6) represents physically the steady-state flow of a fluid that is contained in the cavity shown in Fig. 6 and on which a force acts along CD from C to D.

In general, boundary value problem (4.1) through (4.6) cannot be solved by means of existing analytical techniques. We shall describe, then, a method for approximating the solution. This method was devised by extensive experimentation on a computer, and no rigorous convergence proofs are as yet available. Under such tenuous circumstances, we shall have to show, at least, that our results are consistent with known physical or qualitative mathematical results.

The approach will be to reformulate (4.1) and (4.2) as a double sequence of linear problems, to each of which we then apply some of the ideas and methods discussed at the beginning of Sec. 3. Hence, for a fixed, positive integer n, set $h = 1/n$. Starting at $(0, 0)$ with grid size h, construct R_h and S_h.

For given tolerances ϵ_1 and ϵ_2 let us proceed to construct on R_h a sequence of discrete stream functions

$$(4.7) \qquad \psi^{(0)}, \psi^{(1)}, \psi^{(2)}, \cdots$$

and on $R_h \cup S_h$ a sequence of discrete vorticity functions

$$(4.8) \qquad \omega^{(0)}, \omega^{(1)}, \omega^{(2)}, \cdots$$

such that for some integer k both the following are valid:

$$(4.9) \qquad \left| \psi^{(k)} - \psi^{(k+1)} \right| < \epsilon_1, \qquad \text{on } R_h,$$

$$(4.10) \qquad \left| \omega^{(k)} - \omega^{(k+1)} \right| < \epsilon_2, \qquad \text{on } R_h \cup S_h.$$

Initially, set

$$(4.11) \qquad \psi^{(0)} = C_1, \qquad \text{on } R_h$$

(4.12) $\qquad\qquad \omega^{(0)} = C_2, \qquad$ on $R_h \cup S_h,$

where C_1 and C_2 are constants.

To produce the second iterate $\psi^{(1)}$ of (4.7), proceed as follows. At each point of R_h of the form (h, ih), $i = 2, \cdots, n - 2$, approximate (4.3) by

(4.13) $\qquad\qquad \psi(h, ih) = \dfrac{\psi(2h, ih)}{4}.$

At each point of R_h of the form (ih, h), $i = 1, 2, \cdots, n - 1$, approximate (4.4) by

(4.14) $\qquad\qquad \psi(ih, h) = \dfrac{\psi(ih, 2h)}{4}.$

At each point of R_h of the form $(1 - h, ih)$, $i = 2, 3, \cdots, n - 2$, approximate (4.5) by

(4.15) $\qquad\qquad \psi(1 - h, ih) = \dfrac{\psi(1 - 2h, ih)}{4}.$

At each point of R_h of the form $(ih, 1 - h)$, $i = 1, 2, \cdots, n - 1$, approximate (4.6) by

(4.16) $\qquad\qquad \psi(ih, 1 - h) = \dfrac{h}{2} + \dfrac{\psi(ih, 1 - 2h)}{4}.$

And at each remaining point of R_h write down the difference analogue

(4.17) $\quad -4\psi(x, y) + \psi(x + h, y) + \psi(x, y + h) + \psi(x - h, y)$
$$+ \psi(x, y - h) = -h^2 \omega^{(0)}(x, y)$$

of (4.1). Solve the linear algebraic system generated by (4.13) through (4.17) by iteration and denote this solution by $\bar{\psi}^{(1)}$. Then, on R_h, $\psi^{(1)}$ is defined by the smoothing formula

(4.18) $\qquad \psi^{(1)} = \rho\psi^{(0)} + (1 - \rho)\bar{\psi}^{(1)}, \qquad 0 \leq \rho \leq 1.$

Special formulas (4.13) through (4.16) were found to be essential on the "inner boundary" points, which are crossed in Fig. 7, to achieve numerical convergence to (4.3) through (4.6) for small h, whereas smoothing formula (4.18) was found to be essential for convergence of (4.7) and (4.8) for large R.

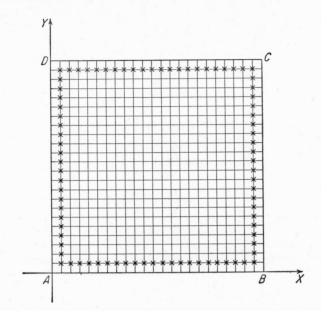

FIG. 7.

To produce the second iterate $\omega^{(1)}$ of sequence (4.8) proceed as follows. At each point of S_h of the form $(ih, 0)$, $i = 0, 1, 2, \cdots, n$, set

$$(4.19) \qquad \bar{\omega}^{(1)}(ih, 0) = -\frac{2\psi^{(1)}(ih, h)}{h^2};$$

at each point of S_h of the form $(0, ih)$, $i = 1, 2, \cdots, n - 1$, set

$$(4.20) \qquad \bar{\omega}^{(1)}(0, ih) = -\frac{2\psi^{(1)}(h, ih)}{h^2};$$

at each point of S_h of the form $(1, ih)$, $i = 1, 2, \cdots, n - 1$, set

$$(4.21) \qquad \bar{\omega}^{(1)}(1, ih) = -\frac{2\psi^{(1)}(1 - h, ih)}{h^2};$$

and at each point of S_h of the form $(ih, 1)$, $i = 0, 1, 2, \cdots, n$, set

$$(4.22) \qquad \bar{\omega}^{(1)}(ih, 1) = \frac{2}{h} - \frac{2\psi^{(1)}(ih, 1 - h)}{h^2}.$$

Next, at each point (x, y) in R_h set

$$\alpha = \psi^{(1)}(x + h, y) - \psi^{(1)}(x - h, y),$$
$$\beta = \psi^{(1)}(x, y + h) - \psi^{(1)}(x, y - h),$$

and approximate (4.2), appropriately, by

(4.23) $\left(-4 - \dfrac{\alpha\mathcal{R}}{2} - \dfrac{\beta\mathcal{R}}{2}\right) \omega(x, y) + \omega(x + h, y)$

$$+ \left(1 + \frac{\alpha\mathcal{R}}{2}\right) \omega(x, y + h)$$

$$+ \left(1 + \frac{\beta\mathcal{R}}{2}\right) \omega(x - h, y) + \omega(x, y - h)$$

$$= 0; \quad \text{if} \quad \alpha \geq 0, \quad \beta \geq 0,$$

(4.24) $\left(-4 - \dfrac{\alpha\mathcal{R}}{2} + \dfrac{\beta\mathcal{R}}{2}\right) \omega(x, y) + \left(1 - \dfrac{\beta\mathcal{R}}{2}\right) \omega(x + h, y)$

$$+ \left(1 + \frac{\alpha\mathcal{R}}{2}\right) \omega(x, y + h) + \omega(x - h, y) + \omega(x, y - h)$$

$$= 0; \quad \text{if} \quad \alpha \geq 0, \quad \beta < 0,$$

(4.25) $\left(-4 + \dfrac{\alpha\mathcal{R}}{2} - \dfrac{\beta\mathcal{R}}{2}\right) \omega(x, y) + \omega(x + h, y) + \omega(x, y + h)$

$$+ \left(1 + \frac{\beta\mathcal{R}}{2}\right) \omega(x - h, y) + \left(1 - \frac{\alpha\mathcal{R}}{2}\right) \omega(x, y - h)$$

$$= 0; \quad \text{if} \quad \alpha < 0, \quad \beta \geq 0.$$

(4.26) $\left(-4 + \dfrac{\alpha\mathcal{R}}{2} + \dfrac{\beta\mathcal{R}}{2}\right) \omega(x, y) + \left(1 - \dfrac{\beta\mathcal{R}}{2}\right) \omega(x + h, y)$

$$+ \omega(x, y + h) + \omega(x - h, y) + \left(1 - \frac{\alpha\mathcal{R}}{2}\right) \omega(x, y - h)$$

$$= 0; \quad \text{if} \quad \alpha < 0, \quad \beta < 0.$$

Solve the linear algebraic system generated by (4.23) through (4.26) by iteration and denote the solution by $\bar{\omega}^{(1)}$. Finally, on all of $R_h \cup S_h$, define $\omega^{(1)}$ by the smoothing formula

(4.27) $\qquad \omega^{(1)} = \mu\omega^{(0)} + (1 - \mu)\bar{\omega}^{(1)}, \qquad 0 \leq \mu \leq 1.$

The reason for the four equations (4.23) through (4.26) instead of one is that $\partial\omega/\partial x$ and $\partial\omega/\partial y$ have been approximated by forward, or backward difference approximations in a fashion that would always make the coefficient of $\omega(x, y)$ negative for all $\Re$ [25c]. This device yields diagonal dominance for *all* Reynolds numbers, and, of course, diagonal dominance is a basic ingredient for convergence of iteration [25c].

Proceed next to determine $\psi^{(2)}$ on R_h from $\omega^{(1)}$ and $\psi^{(1)}$ in the same fashion as $\psi^{(1)}$ was determined from $\omega^{(0)}$ and $\psi^{(0)}$. Then construct $\omega^{(2)}$ on $R_h \cup S_h$ from $\omega^{(1)}$ and $\psi^{(2)}$ in the same fashion as $\omega^{(1)}$ was determined from $\omega^{(0)}$ and $\psi^{(1)}$. In the indicated fashion, construct the sequences (3.1) and (3.2). Terminate the computation when (3.3) and (3.4) are valid.

Finally, when $\psi^{(k)}$ and $\omega^{(k)}$ are verified by substitution to be solutions of the difference analogues of (2.1) and (2.2), they are

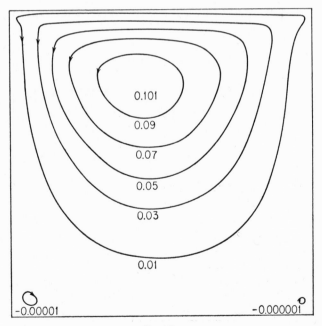

Fig. 8.

taken to be the numerical approximations of $\psi(x, y)$ and $\omega(x, y)$, respectively.

Of the many examples run using the above method (see [25e] for parameter inputs and for other examples), let us discuss only two. In Fig. 8 is shown the graph of ψ for Reynolds number 50 and $h = \frac{1}{40}$. The flow shows one primary vortex and two secondary corner vortices. This flow has been verified experimentally by Pan and Acrivos [44]. In Fig. 9 is shown the double spiral

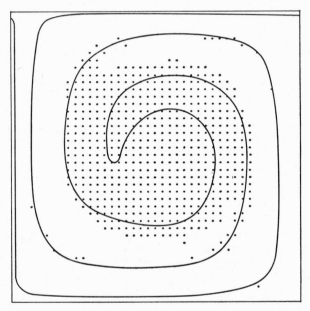

Fig. 9.

graph of $\omega = 1.630$ for Reynolds number 100,000, $h = \frac{1}{40}$. At the dotted points the vorticity is between 1.6 and 1.7. These results are corroborated by Batchelor [4], who proved that vorticity in a large subregion of R converges to a constant as $\Re \to \infty$.

Finally, it should be noted that the ideas of this section apply easily [25d] to nonsteady problems that have steady-state solutions.

REFERENCES AND SOURCES FOR FURTHER READING

1. Alder, B. J., "Studies in molecular dynamics. III. A mixture of hard spheres," *Jour. Chem. Phys.*, **40** (1964), 2724–2730.

2. Ames, W. F., *Nonlinear Partial Differential Equations in Engineering.* New York: Academic Press, 1965.

3. Amsden, A. A., "The particle-in-cell method for the calculation of the dynamics of compressible fluids," Rpt. 3406, Los Alamos Scientific Lab, Los Alamos, N.M., 1966.

4. Batchelor, G. K., "On steady laminar flow with closed streamlines at large Reynolds numbers," *Jour. Fluid Mech.*, **1** (1956), 177–190.

5. Berger, R. L., and N. Davids, "General computer method analysis of condition and diffusion in biological systems with distributive sources," *Rev. Sci. Instr.*, **36** (1965), 88–93.

6. Boughner, R. T., "The discretization error in the finite difference solution of the linearized Navier-Stokes equations for incompressible fluid flow at large Reynolds number," TM 2165, Oak Ridge Nat. Lab., Oak Ridge, Tenn., 1968.

7. Bramble, J. H., "Error estimates for difference methods in forced vibration problems," *SIAM Jour. Num. Anal.*, **3** (1966), 1–12.

8. Chorin, A. J., "The numerical solution of the Navier-Stokes equations for an incompressible fluid," *Bull. AMS*, **73** (1967), 928–931.

9. Collatz, L., *Numerical Treatment of Differential Equations.* Berlin: Springer, 1960.

10. Concus, P., "Numerical solution of the minimal surface equation," *Math. Comp.*, **21** (1967), 340–350.

11. Courant, R., K. Friedrichs, and H. Lewy, "Uber die partiellen Differenzengleichungen der Mathematischen Physik," *Math. Ann.*, **100** (1958), 32–74.

12. Courant, R., E. Isaacson, and M. Rees, "On the solution of nonlinear hyperbolic differential equations by finite differences," *Comm. Pure Appl. Math.*, **5** (1952), 243–255.

13. Cryer, C. W., "Stability analysis in discrete mechanics," Tech. Rpt. 67, Dept. of Computer Science, University of Wisconsin, Madison, Wis., 1969.

14. Danby, J. M. A., *Fundamentals of Celestial Mechanics*. New York: Macmillan, 1962.

15. Davidenko, D. F., "Construction of difference equations for approximating the solution of the Euler-Poisson-Darboux equation" (in Russian), *Dokl. Akad. Nauk SSSR*, **142** (1962), 510–513.

16. Davis, P., and P. Rabinowitz, "Numerical experiments in potential theory using orthonormal functions," *Jour. Wash. Acad. Sciences*, **46** (1956), 12–17.

17. Deeter, C. R., and G. Springer, "Discrete harmonic kernels," *Jour. Math. and Mech.*, **14** (1965), 413–438.

18. Douglas, J., Jr., "A survey of numerical methods for parabolic differential equations," in *Advances in Computers*, F. L. Alt and M. Rubinoff, eds., Vol. II. New York: Academic Press, 1961, pp. 1–55.

19. Dreyfus, S. E., "The numerical solution of nonlinear optimal control problems," in *Numerical Solutions of Nonlinear Differential Equations*, D. Greenspan, ed. New York: Wiley, 1966, pp. 97–113.

20. Duffin, R. J., "Basic properties of discrete analytic functions," *Duke Math. Jour.*, **23** (1956), 335–363.

21. Fermi, E., J. R. Pasta, and S. Ulam, "Studies of nonlinear problems. I," Rpt. #1940, Los Alamos Scientific Lab., Los Alamos, N.M., 1955.

22. Forsythe, G. E., and W. Wasow, *Finite-Difference Methods for Partial Differential Equations*. New York: Wiley, 1960.

23. Fox, L., *Numerical Solution of Ordinary and Partial Differential Equations*. Reading, Mass.: Addison-Wesley, 1962.

24. Fromm, J. E., and F. H. Harlow, "Numerical solution of the problem of vortex street development," *Phys. of Fluids*, **6** (1963), 975–985.

25. Greenspan, D., (a) "Resolution of classical capacity problems by means of a digital computer," *Can. Jour. Phys.*, **44** (1966), 2605–2613.
(b) "On approximating extremals of functionals. Part II. Theory and generalizations related to boundary value problems for nonlinear differential equations," *Int. Jour. Eng. Sci.*, **5** (1967), 571–588.
(c) *Lectures on the Numerical Solution of Linear, Singular, and Nonlinear Differential Equations*. Englewood Cliffs, N.J.: Prentice-Hall, 1968.
(d) "Numerical solution of a class of nonsteady cavity flow problems," *BIT*, **8** (1968), 287–294.

(e) "Numerical studies of prototype cavity flow problems," *The Comp. Jour.*, **12** (1969), 89–94.

(f) "Discrete, nonlinear string vibrations," *The Comp. Jour.*, **13** (1970), 195–201.

(g) "Numerical studies of the 3-body problem," SIAM Jour. Applied Math., **20** (1971), 67–78.

(h) *Introduction to Numerical Analysis and Applications.* Chicago, Ill.: Markham, 1971.

26. Greenspan, D., and P. C. Jain, "Numerical study of subsonic fluid flow by a combination variational integral-finite difference technique," *Jour. Math. Anal. Appl.*, **18** (1967), 85–111.

27. Greenspan, D., and P. Werner, "A numerical method for the exterior Dirichlet problem for the reduced wave equation," *Arch. Rat. Mech. Anal.*, **23** (1966), 288–316.

28. Hamming, R. W., *Numerical Methods for Scientists and Engineers.* New York: McGraw-Hill, 1962.

29. Hansen, K. F., B. V. Koen, and W. W. Little, Jr., "Stable numerical solutions of the reactor kinetics equations," *Nuclear Sci. Eng.*, **22** (1965), 51–59.

30. John, F., "On integration of parabolic equations by difference methods," *Comm. Pure Appl. Math.*, **5** (1952), 155–211.

31. Kawaguti, M., "Numerical solution of the Navier-Stokes equations for the flow in a two-dimensional cavity," *Jour. Phys. Soc. Japan*, **16** (1961), 2307–2315.

32. Keller, H. B., *Numerical Methods for Two-Point Boundary Value Problems.* Waltham, Mass.: Blaisdell, 1968.

33. Kemeny, J. G., and J. L. Snell, *Mathematical Models in the Social Sciences.* New York: Ginn and Co., 1962.

34. Kreiss, H. O., "Difference approximations for the initial-boundary value problem for hyperbolic differential equations," in *Numerical Solutions of Nonlinear Differential Equations.* D. Greenspan, ed. New York: Wiley, 1966, pp. 141–166.

35. Lax, P. D., "Nonlinear partial differential equations and computing," *SIAM Review*, **11** (1969), 7–19.

36. Leith, C. E., "Numerical hydrodynamics of the atmosphere," in *Proc. Symp. Applied Math. of Amer. Math. Soc.* Providence, R. I.: Amer. Math. Soc., 1967, pp. 125–137.

37. May, M. M., and R. H. White, "Hydrodynamic calculation of general-relativistic collapse," *The Physical Review*, **141** (1966), 1232–1241.

38. Mehta, P. K., "Cylindrical and spherical elastoplastic stress waves by a unified direct analysis method," *AIAA Jour.*, **5** (1967), 2242–2248.

39. Melbourne, W. G., "Lunar and planetary flight mechanics," Jet Propulsion Laboratory, Pasadena, Calif., January, 1968.

40. Miller, R. H., and N. Alton, "Three dimensional n-body calculations," ICR Quart. Rpt. #18, University of Chicago, Chicago, Ill., 1968.

41. Miracle, C. L., "Approximate solutions of the telegrapher's equation by difference equation methods," *Jour. SIAM*, **10** (1962), 517–527.

42. von Neumann, J., "Proposal and analysis of a new numerical method for the treatment of hydrodynamical shock problems," in *Collected Works of John von Neumann*, Vol. VI. New York: Pergamon, 1963, pp. 361–379.

43. Noh, W. F., "CEL: a time-dependent, two-space-dimensional, coupled Eulerian-Lagrange code," in *Methods in Computational Physics*. Balder et al., eds. New York: 1964, pp. 117–180.

44. Pan, F., and A. Acrivos, "Steady flows in rectangular cavities," *Jour. Fluid Mech.*, **28** (1967), 643–655.

45. Pearson, C. E., "A computational method for viscous flow problems," *Jour. Fluid Mech.*, **21** (1965), 611–622.

46. Phillips, N. A., "Numerical weather prediction," in *Advances in Computers*, Vol. I. New York: Academic Press, 1960, pp. 43–91.

47. Polya, G., and G. Szego, *Isoperimetric Inequalities in Mathematical Physics*. Princeton, N.J.: Princeton University Press, 1951.

48. Preisendorfer, R. W., *Radiative Transfer on Discrete Spaces*. New York: Pergamon, 1965.

49. Richtmyer, R. D., and K. W. Morton, *Difference Methods for Initial Value Problems*, 2nd ed. New York: Interscience, 1967.

50. Saltzer, C., "Discrete potential and boundary value problems," *Duke Math. Jour.*, **31** (1964), 299–320.

51. Shortley, G. H., R. Weller, P. Darbey, and E. H. Gamble, "Numerical solution of axisymmetrical problems, with applications to electrostatics and torsion," Eng. Exp. Sta. Bull. No. 128, Ohio State University, 1947.

52. Siry, J. W., J. P. Murphy, and I. J. Cole, "The Goddard general orbit

determination system," NASA-TM-X-63413; X-550-68-218, NASA, Goddard Space Flight Ctr., Greenbelt, Md., 1968.

53. Smith, J., "The coupled equation approach to the numerical solution of the biharmonic equation by finite differences. II." SIAM Jour. Num. Anal., 7 (1970), 104–111.

54. Soare, M., *Application of Finite Difference Equations to Shell Analysis.* New York: Pergamon, 1965.

55. Synge, J. L., *The Hypercircle in Mathematical Physics.* Cambridge: Cambridge University Press, 1957.

56. Urabe, M., "Numerical study of periodic solutions of the van der Pol equation," in *International Symposium on Nonlinear Differential Equations and Nonlinear Mechanics.* J. P. LaSalle and S. Lefschetz, eds. New York: Academic Press, 1963, pp. 184–192.

57. Whitrow, G. J., *The Natural Philosophy of Time.* New York: Harper and Row, 1961.

58. Zlamal, M., "On the finite element method," *Num. Mat.,* **12** (1968), 394–409.

SCATTERING THEORY

Tosio Kato

1. INTRODUCTION

It would be impractical to give here a general survey of scattering theory, which has already grown into a large body of mathematical theory with many branches.† The purpose of this article is to illustrate the methods of abstract scattering theory by using a concrete, simple, but nontrivial, example. This theory works within the framework of the theory of linear operators in Hilbert space. In particular we avoid the use of the plane wave solutions of the Schrödinger equation, which have no direct physical meaning.

We consider a classical (= nonrelativistic) quantum mechanical system consisting of a single particle moving in a potential field V. The Schrödinger equation for such a system may be written, with an appropriate choice of units, as

---

† General references for scattering theory are [7, 14, 22, 25, 32]. The time-dependent formalism employed in this article was first given by [13, 19]. For recent results in abstract scattering theory, see, e.g., [3, 18, 26, 29] and the references given there.

(1.1) $i \dfrac{d\psi}{dt} = H_\kappa \psi, \qquad H_\kappa = H_0 + \kappa V,$

where H_κ is the Hamiltonian operator, with

(1.2) $H_0 = -\Delta = -\left(\dfrac{\partial^2}{\partial x^2} + \dfrac{\partial^2}{\partial y^2} + \dfrac{\partial^2}{\partial z^2} \right)$

denoting the kinetic energy. H_0 is also regarded as the unperturbed Hamiltonian. κ is a real parameter, introduced for computational convenience and often assumed to be small to simplify the argument. The potential $V = V(\mathbf{r})$ is a real-valued function; it is *not* assumed to be spherically symmetric. $\mathbf{r}$ stands for (x, y, z).

The solution of (1.1) may be written

(1.3) $\psi(t) = e^{-itH_\kappa}\phi, \qquad \phi = \psi(0).$

Mathematically, H_κ is a *self-adjoint operator* in the basic Hilbert space† $\mathbf{H} = L^2(R^3)$. It is known that this is true if $H_0 = -\Delta$ is taken in a generalized sense and V satisfies certain general conditions.‡ Thus e^{-itH_κ} is a *unitary operator* for each real t and forms a *continuous unitary group*. For each $\phi \in \mathbf{H}$, $\psi(t) \in \mathbf{H}$ is well defined and $||\psi(t)|| = ||\phi||$.

It is usually taken for granted that $\psi(t)$ behaves for large values of $|t|$ as if the particle were free; i.e.,

(1.4) $\psi(t) = e^{-itH_\kappa}\phi \sim e^{-itH_0}\phi_\pm, \qquad t \to \pm \infty,$

at least if ϕ is a "continuum state" of H_κ (orthogonal to all bound states). This is a basic assumption in scattering theory, but we shall prove it under certain assumptions on V.

The unitarity of e^{-itH_κ} and e^{-itH_0} implies that $||\phi|| = ||\phi_+|| = ||\phi_-||$. It follows that $\phi_\pm$ are uniquely determined by ϕ and that the maps§ $\phi \mapsto \phi_\pm$ are *isometric*. It is also assumed (and will be

---

† $L^2(R^3)$ is the Hilbert space consisting of all complex-valued square-integrable functions on the euclidean space R^3, with the inner product $(u, v) = \int u(\mathbf{r})v(\mathbf{r})^* d\mathbf{r}$ (* denotes the complex conjugate) and the norm $||u|| = (u, u)^{1/2}$.

‡ See, e.g., [14], p. 302.

§ $x \mapsto y$ denotes the mapping or function that maps x into y. When we want to emphasize the *function f* as distinguished from its *values* $f(x)$, we write $f : x \mapsto y = f(x)$ or $f : x \mapsto y$ or simply $x \mapsto y$.

proved later) that each of the collections $\{\phi_\pm\}$ exhausts all of **H** when ϕ varies over all the continuum states of H_κ. Thus (1.4) may be interpreted to mean that any "initial free state" ϕ_- goes over to the "final free state" ϕ_+ under the action of the scatterer V. This defines the *scattering operator* S by

$$(1.5) \qquad\qquad \phi_+ = S\phi_-.$$

Under the above assumptions, S is a *unitary* operator. All the information about the scattering of the particle under consideration should be contained in S.

For mathematical purposes, it is convenient to introduce the *wave operators* $W_\pm$ defined by

$$(1.6) \qquad\qquad \phi = W_+\phi_+ = W_-\phi_-.$$

Then

$$(1.7) \qquad\qquad S = W_+^{-1}W_-.$$

$W_\pm$ are *isometric* but in general not unitary, for their ranges are the set of continuum states of H_κ, which need not be the whole of **H**.

Equation (1.4) can now be rewritten as $e^{itH_\kappa}e^{-itH_0}\phi_\pm \to \phi = W_\pm\phi_\pm$, $t \to \pm\infty$. Thus

$$(1.8) \qquad\qquad W_\pm = \text{s-}\lim_{t\to\pm\infty} e^{itH_\kappa}e^{-itH_0},$$

where s-lim means strong limit in **H**.

From (1.8) one sees easily that $W_\pm$ have the *intertwining property*:

$$(1.9) \qquad\qquad e^{itH_\kappa}W_\pm = W_\pm e^{itH_0}, \qquad -\infty < t < \infty.$$

Then (1.7) gives

$$(1.10) \qquad\qquad e^{itH_0}S = Se^{itH_0}, \qquad -\infty < t < \infty.$$

This implies that S commutes with H_0, so that S is *diagonalized* simultaneously with H_0. More precisely, let us introduce the spectral representation for H_0 furnished by the Fourier transform

$$(1.11) \qquad\qquad \hat{\phi}(\mathbf{k}) = (2\pi)^{-3/2} \int e^{-i\mathbf{k}\cdot\mathbf{r}}\phi(\mathbf{r})\, d\mathbf{r}.$$

H_0 is diagonal in this representation in the sense that

(1.12) $\qquad (H_0\hat{\phi})(\mathbf{k}) = |\mathbf{k}|^2\hat{\phi}(\mathbf{k}).$

Set

(1.13) $\qquad \phi(\lambda;\boldsymbol{\omega}) = 2^{-1/2}|\mathbf{k}|^{1/2}\hat{\phi}(\mathbf{k}),$

where

(1.14) $\qquad \lambda = |\mathbf{k}|^2, \qquad \boldsymbol{\omega} = \dfrac{\mathbf{k}}{|\mathbf{k}|}.$

λ is the kinetic energy, and $\boldsymbol{\omega}$ is the unit vector in the direction of the momentum $\mathbf{k}$. The factor in (1.13) is so chosen that

(1.15) $\quad ||\phi||^2 = ||\hat{\phi}||^2 = \displaystyle\int |\hat{\phi}(\mathbf{k})|^2 |\mathbf{k}|^2 \, d|\mathbf{k}| \, d\boldsymbol{\omega} =$
$$\int d\lambda \int |\phi(\lambda;\boldsymbol{\omega})|^2 \, d\boldsymbol{\omega}.$$

In this notation, the inverse Fourier transform formula becomes

(1.16) $\qquad \phi(\mathbf{r}) = (2\pi)^{-3/2} \displaystyle\int e^{i\mathbf{k}\cdot\mathbf{r}}\hat{\phi}(\mathbf{k}) \, d\mathbf{k}$
$$= \left(\frac{1}{4\pi^{3/2}}\right) \int e^{i\lambda^{1/2}\boldsymbol{\omega}\cdot\mathbf{r}}\phi(\lambda;\boldsymbol{\omega})\lambda^{1/4} \, d\lambda \, d\boldsymbol{\omega}.$$

Let Ω denote (the surface of) the unit sphere. For almost every fixed λ, $\boldsymbol{\omega} \mapsto \phi(\lambda;\boldsymbol{\omega})$ is a square integrable function of $\boldsymbol{\omega} \in \Omega$; we denote this function by $\phi(\lambda)$ so that $\phi(\lambda) \in \mathbf{P} \equiv L^2(\Omega)$. The state $\phi \in \mathbf{H}$ can now be represented by the function $\phi : \lambda \mapsto \phi(\lambda) \in \mathbf{P}$. Then (1.15) means that

(1.17) $\qquad ||\phi||^2 = \displaystyle\int_0^\infty ||\phi(\lambda)||^2 \, d\lambda.$

Note that the norm $||\phi||$ on the left of (1.17) is that of $\mathbf{H}$, but $||\phi(\lambda)||$ on the right is the norm of $\mathbf{P}$.

We shall express this relation by

(1.18) $\qquad \phi \leftrightarrow \{\phi(\lambda)\}.$

According to (1.12), we have

(1.19) $\qquad H_0\phi \leftrightarrow \{\lambda\phi(\lambda)\}.$

In this sense (1.18) is a *spectral representation* for H_0. Mathematically, (1.18) means that $\mathbf{H}$ may be identified with $L^2(0,\infty;\mathbf{P})$ [L^2-space of $\mathbf{P}$-valued functions on the interval $(0,\infty)$].

Since S commutes with H_0, it does not mix different components in this representation. In other words, S acts on each component $\phi(\lambda)$ separately:

(1.20) $S\phi \leftrightarrow \{S(\lambda)\phi(\lambda)\}.$

(We do not give a general proof of this, but it will be proved later in our special case.) In (1.20), $S(\lambda)$ is for each λ a unitary operator in $\mathbf{P}$, since S is unitary in $\mathbf{H}$. In this sense S is the *direct integral* of unitary operators $S(\lambda)$. $S(\lambda)$ corresponds to the S-matrix in the physical literature. The basic problem in scattering theory is to compute $S(\lambda)$ for all $\lambda \geq 0$; then such quantities as differential and total scattering cross sections should follow automatically. We shall do this for our example in the following sections.

2. ASSUMPTIONS AND BASIC LEMMAS

We shall show that the program of Sec. 1 can be carried out rigorously under the assumptions[†]

(2.1) $V \in L^{3/2}(R^3), \qquad V \in L^{\infty}(R^3),$

provided $|\kappa|$ is sufficiently small.

Several comments are in order: (1) The assumption $V \in L^{\infty}(R^3)$ is not necessary; it is made here only to avoid inessential complications in the argument.[‡] Even the first condition of (2.1) is not necessary; it may be replaced by the weaker condition that[§] $\iint |V(\mathbf{r})||V(\mathbf{r}')||\mathbf{r} - \mathbf{r}'|^{-2}\, d\mathbf{r}\, d\mathbf{r}' < \infty$. (2) The assumption that $|\kappa|$ be small can also be eliminated. But to do so one has to use

---

[†] For $s \geq 1$, $L^s(E)$ denotes the Banach space consisting of all complex-valued functions v on a measure space E such that $||v||_{L^s} = (\int_E |v(x)|^s\, dx)^{1/s}$ is finite. When $s = \infty$, this norm is to be replaced by ess sup $|v| = $ the smallest number M such that $|v| \leq M$ almost everywhere.

[‡] For the proof without this assumption, see [15].

[§] This is called the Rollnik condition in [29]. That this condition is sufficient for our proofs is easily seen, except possibly the proof in Lemma 5 that $F_A(\lambda)$ is continuous in λ in the B_4-norm. But this can be done easily by approximating V with bounded functions with compact supports.

more complicated methods than those given below.† (3) If we disregard the second condition of (2.1) according to the remark (1), then (2.1) implies that V may have singularities of the form $|\mathbf{r} - \mathbf{r}_k|^{-2+\epsilon}$ at isolated points $\mathbf{r}_k$, $k = 1, 2, \cdots$, and that V may decrease at infinity as slowly as $|\mathbf{r}|^{-2-\epsilon}$, where $\epsilon > 0$. Actually, it is known that a decay rate at infinity of the order $|\mathbf{r}|^{-1-\epsilon}$ is sufficient for most purposes,‡ but we do not prove it here.

We start with several simple consequences of the assumptions (2.1). We factor V into the form§

$$(2.2) \qquad V(\mathbf{r}) = A(\mathbf{r})B(\mathbf{r})$$

in such a way that A and B are real functions with

$$(2.3) \qquad A, B \in L^3(R^3), \qquad A, B \in L^\infty(R^3).$$

For example, one may choose $A(\mathbf{r}) = |V(\mathbf{r})|^{1/2}$.

For each complex number ζ that does not lie on the nonnegative real axis, the *resolvent*

$$(2.4) \qquad R_0(\zeta) = (H_0 - \zeta)^{-1}$$

of H_0 is defined and is an integral operator with kernel (Green's function)

$$(2.5) \qquad g_0(\mathbf{r}, \mathbf{r}'; \zeta) = \frac{e^{i\zeta^{1/2}|\mathbf{r}-\mathbf{r}'|}}{4\pi|\mathbf{r} - \mathbf{r}'|},$$

where $\zeta^{1/2}$ is to be chosen so that $\operatorname{Im} \zeta^{1/2} > 0$. We have

$$(2.6) \qquad |g_0(\mathbf{r}, \mathbf{r}'; \zeta)| \leq \frac{1}{4\pi|\mathbf{r} - \mathbf{r}'|}.$$

---

† In this sense this article is essentially a *small perturbation theory*. For the case of large $|\kappa|$, see, e.g., [18, 20].

‡ See [17, 21].

§ This factorization is a well-known technique to construct a nice operator $Q_0(\zeta)$ by (2.7), which gives a convenient form to the Neumann (or Born) expansion (2.16) of the resolvent $R_\kappa(\zeta)$. Since $Q_0(\zeta)$ is an operator of Hilbert-Schmidt type, all the following arguments proceed smoothly within the basic Hilbert space. In a certain sense, however, this is a tricky device. The use of factorization may be avoided, but then one would have to introduce a new topology in the space of operators considered; see [10, 18, 26].

Set

(2.7) $$Q_0(\zeta) = BR_0(\zeta)A,$$

where A and B are regarded as operators of multiplication.

LEMMA 1: $Q_0(\zeta)$ *is an operator of Hilbert-Schmidt type. If we denote by* $|| \ ||_2$ *the Hilbert-Schmidt norm, we have*

(2.8) $$||Q_0(\zeta)||_2 \le c||A||_{L^3}||B||_{L^3} \equiv M,$$

where c is a numerical constant. $Q_0(\zeta)$ depends on ζ continuously in the sense of $|| \ ||_2$ *(i.e.,* $||Q_0(\zeta) - Q_0(\zeta_0)||_2 \to 0$ *as $\zeta \to \zeta_0$). Similar results hold for $AR_0(\zeta)A$ and $BR_0(\zeta)B$ instead of $Q_0(\zeta)$.*

Proof: $Q_0(\zeta)$ is an integral operator with kernel

(2.9) $$B(\mathbf{r})g_0(\mathbf{r}, \mathbf{r}'; \zeta)A(\mathbf{r}'),$$

which is majorized by [see (2.6)]

(2.10) $$\frac{|B(\mathbf{r})||A(\mathbf{r}')|}{4\pi|\mathbf{r} - \mathbf{r}'|}.$$

It follows from (2.3) that the square integral of (2.10) on $R^3 \times R^3$ if finite, not exceeding a numerical constant times $||A||_{L^3}^2||B||_{L^3}^2$; this is due to the Sobolev inequality.† Hence (2.9) is of Hilbert-Schmidt type and satisfies (2.8). To prove that $||Q_0(\zeta) - Q_0(\zeta_0)||_2 \to 0$, it suffices to show that

$$\iint |B(\mathbf{r})[g_0(\mathbf{r}, \mathbf{r}'; \zeta) - g_0(\mathbf{r}, \mathbf{r}'; \zeta_0)]A(\mathbf{r}')|^2 \, d\mathbf{r} \, d\mathbf{r}' \to 0.$$

This is easily done with the aid of Lebesgue's dominated convergence theorem. One can handle $AR_0(\zeta)A$ and $BR_0(\zeta)B$ similarly.

LEMMA 2: A, B *are H_0-smooth, by which we mean that for each $u \in \mathbf{H}$ and $\epsilon > 0$*

(2.11) $$\int_{-\infty}^{\infty} [||AR_0(\lambda + i\epsilon)u||^2 + ||AR_0(\lambda - i\epsilon)u||^2] \, d\lambda$$

$$= \int_{-\infty}^{\infty} ||AR_0(\lambda + i\epsilon)u - AR_0(\lambda - i\epsilon)u||^2 \, d\lambda$$

$$\le 4\pi M'||u||^2, \qquad M' = c||A||_{L^3}^2,$$

and similarly for B with constant $M'' = c||B||_{L^3}^2$.

---

† See [30], p. 43.

Proof: The first two expressions in (2.11) are equal because

$$\int_{-\infty}^{\infty} (AR_0(\lambda + i\epsilon)u, AR_0(\lambda - i\epsilon)u) \, d\lambda$$
$$= \int_{-\infty}^{\infty} (R_0(\lambda + i\epsilon)A^2R_0(\lambda + i\epsilon)u, u) \, d\lambda = 0$$

by Cauchy's theorem,† since $R_0(\zeta)A^2R_0(\zeta)$ is analytic in ζ for Im $\zeta \gtrless 0$. To prove the inequality in (2.11), we note that

(2.12) $R_0(\lambda + i\epsilon) - R_0(\lambda - i\epsilon) = 2i\epsilon R_0(\lambda + i\epsilon)R_0(\lambda - i\epsilon).$

Hence

$2\epsilon||AR_0(\lambda + i\epsilon)R_0(\lambda - i\epsilon)A||$
$= ||AR_0(\lambda + i\epsilon)A - AR_0(\lambda - i\epsilon)A|| \leq 2M'$

by Lemma 1 (note that $|| \, || \leq || \, ||_2$). But

$$||AR_0(\lambda + i\epsilon)R_0(\lambda - i\epsilon)A|| = ||AR_0(\lambda + i\epsilon)||^2,$$

for $||TT^*|| = ||T||^2$ is true for any bounded operator T (T^* denotes the adjoint operator to T). Hence

(2.13) $||AR_0(\lambda + i\epsilon)|| \leq (M'/\epsilon)^{1/2}.$

Using (2.12) again, we thus obtain

(2.14) $||AR_0(\lambda + i\epsilon)u - AR_0(\lambda - i\epsilon)u||$
$= 2\epsilon||AR_0(\lambda + i\epsilon)R_0(\lambda - i\epsilon)u||$
$\leq 2\epsilon||AR_0(\lambda + i\epsilon)|| \, ||R_0(\lambda - i\epsilon)u||$
$\leq 2(\epsilon M')^{1/2}||R_0(\lambda - i\epsilon)u||.$

Hence

$$\int_{-\infty}^{\infty} ||AR_0(\lambda + i\epsilon)u - AR_0(\lambda - i\epsilon)u||^2 \, d\lambda$$
$$\leq 4\epsilon M' \int_{-\infty}^{\infty} ||R_0(\lambda - i\epsilon)u||^2 \, d\lambda$$
$$= 4\epsilon M' \int_{-\infty}^{\infty} ([(H_0 - \lambda)^2 + \epsilon^2]^{-1}u, u) \, d\lambda = 4\pi M'||u||^2,$$

where we used the basic property of the resolvent.

Remark: Lemma 2 shows that for each $u \in \mathbf{H}$, the $\mathbf{H}$-valued function $\zeta \mapsto AR_0(\zeta)u$ belongs to the *Hardy class* in each of the

---

† To be rigorous, one has to estimate the function for large values of $|\zeta|$.

upper and lower half-planes. Consequently, its boundary values on the two edges of the real axis exist, in the sense of L^2-convergence as well as almost everywhere pointwise convergence.[†] We denote these boundary values *symbolically* by $AR_0(\lambda \pm i0)u$; note that $R_0(\lambda \pm i0)$ do not make sense.

We now consider the (perturbed) Hamiltonian

$$(2.15) \qquad H_\kappa = H_0 + \kappa V = H_0 + \kappa AB.$$

We note that H_κ is self-adjoint with the same domain as H_0, since V is a bounded operator. The resolvent of H_κ is given by

$$(2.16) \quad R_\kappa(\zeta) = (H_\kappa - \zeta)^{-1} = (H_0 - \zeta + \kappa AB)^{-1}$$
$$= R_0(\zeta) - \kappa R_0(\zeta)ABR_0(\zeta) + \kappa^2 R_0(\zeta)ABR_0(\zeta)ABR_0(\zeta) - \cdots$$
$$= R_0(\zeta) - \kappa R_0(\zeta)A[1 + \kappa Q_0(\zeta)]^{-1}BR_0(\zeta),$$

where we assume that

$$(2.17) \qquad |\kappa| < 1/M$$

so that the series converges by (2.8) and

$$(2.18) \qquad \|[1 + \kappa Q_0(\zeta)]^{-1}\| \leq (1 - M|\kappa|)^{-1}.$$

It follows from (2.16) that

$$(2.19) \qquad BR_\kappa(\zeta) = [1 + \kappa Q_0(\zeta)]^{-1}BR_0(\zeta)$$

and, similarly, $R_\kappa(\zeta)A = R_0(\zeta)A[1 + \kappa Q_0(\zeta)]^{-1}$ or

$$(2.20) \qquad AR_\kappa(\zeta^*) = [1 + \kappa Q_0(\zeta)^*]^{-1}AR_0(\zeta^*).$$

In view of (2.18), we have thus proved

LEMMA 3: *A, B are H_κ-smooth, in the sense stated in Lemma 2, with the constants M', M'' replaced by $M'(1 - M|\kappa|)^{-2}$, $M''(1 - M|\kappa|)^{-2}$, respectively.*

We now consider the unitary groups e^{-itH_0} and e^{-itH_κ}, $-\infty < t < \infty$, which are well defined since H_0 and H_κ are self-adjoint.

LEMMA 4: *For each $u \in \mathbf{H}$ we have*

[†] See [8], p. 55.

$$(2.21) \qquad \int_{-\infty}^{\infty} ||Ae^{-itH_0}u||^2\, dt \leq 2M'||u||^2,$$

$$(2.22) \qquad \int_{-\infty}^{\infty} ||Ae^{-itH_\kappa}u||^2\, dt \leq 2M'||u||^2/(1 - M|\kappa|)^2,$$

and similarly for A, M' replaced by B, M'', respectively.

Proof: Let $\chi_\pm$ be the characteristic functions of the intervals $(0, \infty)$ and $(-\infty, 0)$, respectively. The **H**-valued function $\lambda \mapsto AR_0(\lambda \pm i\epsilon)u$ is the inverse Fourier-Plancherel transform of the function $t \mapsto \pm i(2\pi)^{1/2}\chi_\pm(t)e^{-\epsilon|t|}Ae^{-itH_0}u$, as is easily verified. Thus we have, by Parseval's theorem,

$$2\pi \int_{-\infty}^{\infty} \chi_\pm(t)e^{-2\epsilon|t|}||Ae^{-itH_0}u||^2\, dt = \int_{-\infty}^{\infty} ||AR_0(\lambda \pm i\epsilon)u||^2\, d\lambda.$$

Adding these two equalities, using (2.11), and going to the limit as $\epsilon \to 0$, we obtain (2.21). Similarly, (2.22) can be proved by using Lemma 3.

Remark: Once we know that (2.21) is true, we see that the **H**-valued function $t \mapsto Ae^{-itH_0}u$ belongs to $L^2(-\infty, \infty\,;\mathbf{H})$. Again going to the limit as $\epsilon \to 0$, it follows that the functions

$$\lambda \mapsto AR_0(\lambda \pm i0)u$$

are the inverse Fourier transforms of $t \mapsto \pm i(2\pi)^{1/2}\chi_\pm(t)Ae^{-itH_0}u$. Consequently, the inverse Fourier transform of $t \mapsto Ae^{-itH_0}u$ *is*

$$\lambda \mapsto -i(2\pi)^{-1/2}[AR_0(\lambda + i0)u - AR_0(\lambda - i0)u]$$
$$= (2\pi)^{1/2}AE_0'(\lambda)u,$$

where

$$(2.23) \quad E_0'(\lambda) = (2\pi i)^{-1}[R_0(\lambda + i0) - R_0(\lambda - i0)] = \delta(H_0 - \lambda)$$

symbolically (δ is the delta function). $E_0'(\lambda)$ has no meaning as an operator in **H**, but $\lambda \mapsto AE_0'(\lambda)u$ is well defined as a function in $L^2(-\infty, \infty\,;\mathbf{H})$.

3. THE WAVE AND SCATTERING OPERATORS

THEOREM 1: *The wave operators $W_\pm$ and the scattering operator $S = W_+^{-1}W_-$ exist. They are unitary and depend analytically on κ.*

Proof: We start from the identity

$$(d/dt)e^{itH_\kappa}e^{-itH_0}u = ie^{itH_\kappa}(H_\kappa - H_0)e^{-itH_0}u$$
$$= i\kappa e^{itH_\kappa}ABe^{-itH_0}u,$$

which is valid for all t if u is in the domain of H_0. This gives, after integration,

$$(3.1) \quad e^{it''H_\kappa}e^{-it''H_0}u - e^{it'H_\kappa}e^{-it'H_0}u$$
$$= i\kappa \int_{t'}^{t''} e^{itH_\kappa}ABe^{-itH_0}u \, dt, \qquad t' < t''.$$

Note that (3.1) is true for *all* $u \in \mathbf{H}$, for all the operators involved are bounded.

Let us estimate the last integral. For each $u, v \in \mathbf{H}$,

$$\left| \left(\int_{t'}^{t''} e^{itH_\kappa}ABe^{-itH_0}u \, dt, v \right) \right|$$
$$= \left| \int_{t'}^{t''} (Be^{-itH_0}u, Ae^{-itH_\kappa}v) \, dt \right|$$
$$\leq \int_{t'}^{t''} ||Be^{-itH_0}u|| \, ||Ae^{-itH_\kappa}v|| \, dt$$
$$\leq \left(\int_{t'}^{t''} ||Be^{-itH_0}u||^2 \, dt \right)^{1/2} \left(\int_{t'}^{t''} ||Ae^{-itH_\kappa}v||^2 \, dt \right)^{1/2}$$
$$\leq ||v||(2M')^{1/2}(1 - M|\kappa|)^{-1} \left(\int_{t'}^{t''} ||Be^{-itH_0}u||^2 \, dt \right)^{1/2},$$

where we used the Schwarz inequality and Lemma 4. Since this is true for every $v \in \mathbf{H}$, we have

$$(3.2) \quad \left\| \int_{t'}^{t''} e^{itH_\kappa}ABe^{-itH_0}u \, dt \right\|$$
$$\leq (2M')^{1/2}(1 - M|\kappa|)^{-1} \left(\int_{t'}^{t''} ||Be^{-itH_0}u||^2 \, dt \right)^{1/2}.$$

But since the integral on the right exists when $t'' = \infty$ (see Lemma 4), (3.2) tends to zero as $t', t'' \to \infty$. It follows from (3.1) that $\lim_{t \to \infty} e^{itH_\kappa} e^{-itH_0}u$ exists. Since this is true for every $u \in \mathbf{H}$, $W_+ =$ s-$\lim_{t \to \infty} e^{itH_\kappa} e^{-itH_0}$ exists. Similarly, one proves that W_- exists.

We know that $W_\pm$ are isometric (see Sec. 1). But we shall now show that $W_\pm$ are unitary. To this end we note that the same argument as given above leads to the existence of $Z_\pm = \underset{t \to \pm\infty}{\text{s-lim}}$ $e^{itH_0}e^{-itH_\kappa}$. Then $W_\pm Z_\pm = \underset{t \to \pm\infty}{\text{s-lim}} (e^{itH_\kappa} e^{-itH_0})(e^{itH_0} e^{-itH_\kappa}) = 1$. This implies that the ranges of $W_\pm$ are all of $\mathbf{H}$. Hence $W_\pm$ are unitary, and the same is true of $S = W_+^{-1}W_-$.

We omit the proof of the analytic dependence of $W_\pm$ on κ, which is not difficult. The same for S will be proved later.

Remark: The intertwining relation (1.9) may now be written $e^{itH_\kappa} = W_\pm e^{itH_0}W_\pm^{-1}$, which is equivalent to

$$(3.3) \qquad\qquad H_\kappa = W_\pm H_0 W_\pm^{-1}.$$

Thus H_κ is *unitarily equivalent* to H_0; H_κ has inner structure identical with that of H_0. In particular, H_κ has the same spectrum as H_0, consisting of a pure continuous spectrum extending over the nonnegative real axis. This is a rather special situation due to the assumption that $|\kappa|$ is small.†

THEOREM 2: *S is given by*

$$(3.4) \quad S = 1 - 2i\pi\kappa \int_0^\infty E_0'(\lambda)A[1 + \kappa Q_0(\lambda + i0)]^{-1}BE_0'(\lambda)\, d\lambda.$$

Remark: Equation (3.4) is a symbolic notation; it actually means that for each $u, v \in \mathbf{H}$,

$$(3.5) \quad ((S - 1)u, v)$$

$$= -2i\pi\kappa \int_0^\infty ([1 + \kappa Q_0(\lambda + i0)]^{-1}BE_0'(\lambda)u, AE_0'(\lambda)v)\, d\lambda,$$

where $BE_0'(\lambda)u$ and $AE_0'(\lambda)v$ are functions in $L^2(0, \infty; \mathbf{H})$ (see the Remark at the end of Sec. 2).

Proof of Theorem 2. On exchanging H_0 and H_κ in (3.1) (which entails change of sign) and letting $t' \to -\infty$, $t'' \to \infty$, we obtain

$$(3.6) \qquad Z_+ - Z_- = -i\kappa \int_{-\infty}^\infty e^{itH_0}ABe^{-itH_\kappa}\, dt,$$

the integral converging strongly. Multiplying (3.6) from the right by W_-, using $Z_+W_- = S$, $Z_-W_- = 1$, and (3.3), we obtain

---

† For example, bound states with negative eigenvalues may appear for large $|\kappa|$.

$$(3.7) \qquad S - 1 = -i\kappa \int_{-\infty}^{\infty} e^{itH_0} ABW_- e^{-itH_0} \, dt.$$

Going over to the Fourier transform, and noting the Remark at the end of Sec. 2, we have, by Parseval's theorem,

$$(3.8) \qquad S - 1 = -2i\pi\kappa \int_{-\infty}^{\infty} E_0'(\lambda) ABW_- E_0'(\lambda) \, d\lambda,$$

which is a symbolic formula to be interpreted according to the Remark above. (Note that BW_- is H_0-smooth because B is H_κ-smooth and (3.3) is true.)

We have to dispose of the unknown quantity W_- in (3.8). To this end we return to (3.1), in which we let $t' \to -\infty$, $t'' = 0$, replace u by $e^{-isH_0}u$, and multiply on the left by B. The result is

$$Be^{-isH_0}u - BW_- e^{-isH_0}u$$

$$= i\kappa \int_{-\infty}^{0} Be^{itH_\kappa} ABe^{-i(t+s)H_0} u \, dt$$

$$= i\kappa \int_{-\infty}^{\infty} \chi_+(s - t)Be^{-(s-t)H_\kappa} ABe^{-itH_0} u \, dt.$$

In the Remark at the end of Sec. 2 the (inverse) Fourier transforms of all the functions involved are given. Noting that convolution goes over into $(2\pi)^{1/2}$ times multiplication under the Fourier transform, we obtain

$$(3.9) \qquad BW_- E_0'(\lambda) = [1 - \kappa BR_\kappa(\lambda + i0)A]BE_0'(\lambda),$$

again in a symbolic sense. But $1 - \kappa BR_\kappa(\zeta)A = [1 + \kappa Q_0(\zeta)]^{-1}$, as is easily seen from (2.16). Thus we arrive at (3.4), noting that $E_0'(\lambda) = 0$ for $\lambda < 0$.

Remark: Equation (3.4) can be expanded into a power series in κ:

(3.10)

$$S = 1 + 2i\pi \sum_{n=1}^{\infty} (-\kappa)^n \int_0^{\infty} E_0'(\lambda) AQ_0(\lambda + i0)^{n-1} BE_0'(\lambda) \, d\lambda$$

$$= 1 + 2i\pi \sum_{n=1}^{\infty} (-\kappa)^n \int_0^{\infty} E_0'(\lambda) V[R_0(\lambda + i0)V]^{n-1} E_0'(\lambda) \, d\lambda,$$

which converges for (2.17). Equations (3.4) and (3.10) are convenient expressions for S as an operator in **H**. It is more important, however, to deduce an expression for the S-matrix $S(\lambda)$, which is the object of the following sections.

4. SOME RESULTS ON COMPACT OPERATORS†

To deal with $S(\lambda)$, we need the properties of certain compact (completely continuous) operators acting between different Hilbert spaces. In what follows we denote by $\mathbf{H}$, $\mathbf{K}$, $\mathbf{L}$, $\cdots$ separable Hilbert spaces. We denote by $B(\mathbf{H}, \mathbf{K})$ the set of all bounded linear operators on $\mathbf{H}$ into $\mathbf{K}$. The norm of $T \in B(\mathbf{H}, \mathbf{K})$ is defined by $||T|| = \sup_{||u|| \leq 1} ||Tu||$; note that here u varies over all vectors of $\mathbf{H}$ with $||u|| \leq 1$, whereas $||Tu||$ is the norm of $Tu \in \mathbf{K}$. $B(\mathbf{H}, \mathbf{K})$ is a Banach space under this norm. In particular, the triangle inequality $||T + S|| \leq ||T|| + ||S||$ holds. We write $B(\mathbf{H})$ for $B(\mathbf{H}, \mathbf{H})$. If $T \in B(\mathbf{H}, \mathbf{K})$ and $S \in B(\mathbf{K}, \mathbf{L})$, then $ST \in B(\mathbf{H}, \mathbf{L})$ with $||ST|| \leq ||S|| \, ||T||$.

The adjoint T^* for $T \in B(\mathbf{H}, \mathbf{K})$ belongs to $B(\mathbf{K}, \mathbf{H})$ and is characterized by the property that $(T^*u, v) = (u, Tv)$ for all $u \in \mathbf{K}$ and $v \in \mathbf{H}$, where $(\ ,\ )$ on the left is the inner product in $\mathbf{H}$ and that on the right is in $\mathbf{K}$. We have $||T^*|| = ||T||$.

We denote by $B_2(\mathbf{H}, \mathbf{K})$ [or by $B_2(\mathbf{H})$ if $\mathbf{K} = \mathbf{H}$] the set of all Hilbert-Schmidt operators $T \in B(\mathbf{H}, \mathbf{K})$. These are characterized by the condition that $||T||_2^2 = \Sigma ||Te_n||^2$ is finite, where $\{e_n\}$ is a complete orthonormal set in $\mathbf{H}$. It is known that this sum is independent of the choice of $\{e_n\}$. It is easy to see that $||T||_2 \geq ||T||$ and that the norm $||\ ||_2$ satisfies the triangle inequality. $B_2(\mathbf{H}, \mathbf{K})$ is a Banach space under the norm $||\ ||_2$.

All operators $T \in B_2(\mathbf{H}, \mathbf{K})$ are compact. If in particular $\mathbf{K} = \mathbf{H}$ and T is symmetric ($T^* = T$), T has a pure point spectrum consisting of eigenvalues λ_1, λ_2, $\cdots$ such that $\lambda_n \to 0$ as $n \to \infty$.

If $T \in B_2(\mathbf{H}, \mathbf{K})$ and $S \in B(\mathbf{K}, \mathbf{L})$ or if $T \in B(\mathbf{H}, \mathbf{K})$ and $S \in B_2(\mathbf{K}, \mathbf{L})$, then $ST \in B_2(\mathbf{H}, \mathbf{L})$ and

$$(4.1) \qquad ||ST||_2 \leq ||S|| \, ||T||_2 \quad \text{or} \quad ||ST||_2 \leq ||S||_2 ||T||,$$

---

† For this section, see [5], pp. 1088ff. and [28].

respectively. Again, $T \in B_2(\mathbf{H}, \mathbf{K})$ if and only if $T^* \in B_2(\mathbf{K}, \mathbf{H})$, with

(4.2) $$||T^*||_2 = ||T||_2.$$

If $\mathbf{H} = L^2(E)$ and $\mathbf{K} = L^2(F)$, where E and F are certain measure spaces, $T \in B_2(\mathbf{H}, \mathbf{K})$ if and only if T is an integral operator with the kernel $t(p, q)$ of the Hilbert-Schmidt type. In this case we have

(4.3) $$||T||_2^2 = \iint\limits_{E \times F} |t(p, q)|^2 \, dp \, dq.$$

We need another class $B_4(\mathbf{H}, \mathbf{K})$. It is the set of all $T \in B(\mathbf{H}, \mathbf{K})$ such that $T^*T \in B_2(\mathbf{H})$. We have $B_2(\mathbf{H}, \mathbf{K}) \subset B_4(\mathbf{H}, \mathbf{K})$, as is seen from (4.1). We define the norm $||T||_4$ for $T \in B_4(\mathbf{H}, \mathbf{K})$ by

(4.4) $$||T||_4 = ||T^*T||_2^{1/2}.$$

It follows from (4.1) that

(4.5) $\quad ||T|| \leq ||T||_4 \quad$ for $T \in B_4, \qquad ||T||_4 \leq ||T||_2 \quad$ for $T \in B_2.$

$B_4(\mathbf{H}, \mathbf{K})$ is again a Banach space under the norm $|| \ ||_4$; in particular, $|| \ ||_4$ satisfies the triangle inequality, as one can verify easily.

If $T \in B_4(\mathbf{H}, \mathbf{K})$ and $S \in B(\mathbf{K}, \mathbf{L})$ or if $T \in B(\mathbf{H}, \mathbf{K})$ and $S \in B_4(\mathbf{K}, \mathbf{L})$, then $ST \in B_4(\mathbf{H}, \mathbf{L})$ and

(4.6) $$||ST||_4 \leq ||S|| \ ||T||_4 \quad \text{or} \quad ||ST||_4 \leq ||S||_4||T||,$$

respectively. If $T \in B_4(\mathbf{H}, \mathbf{K})$ and $S \in B_4(\mathbf{K}, \mathbf{L})$, then $ST \in B_2(\mathbf{H}, \mathbf{L})$ with

(4.7) $$||ST||_2 \leq ||S||_4||T||_4.$$

Again, $T \in B_4(\mathbf{H}, \mathbf{K})$ if and only if $T^* \in B_4(\mathbf{K}, \mathbf{H})$, with

(4.8) $$||T^*||_4 = ||T||_4.$$

The following continuity properties are simple consequences of these inequalities. Let T_n, $T \in B_4(\mathbf{H}, \mathbf{K})$, S_n, $S \in B_4(\mathbf{K}, \mathbf{L})$, R_n, $R \in B(\mathbf{K}, \mathbf{L})$ and let $||T_n - T||_4 \to 0$, $||S_n - S||_4 \to 0$, $||R_n - R|| \to 0$. Then $||S_nT_n - ST||_2 \to 0$ and $||R_nT_n - RT||_4 \to 0$.

5. THE S-MATRIX

Let us now return to our problem. We recall that there is a spectral representation $u \leftrightarrow \{u(\lambda)\}$ for each $u \in \mathbf{H} = L^2(R^3)$, where $u(\lambda) \in \mathbf{P} = L^2(\Omega)$ for a. e. $\lambda \geq 0$ [see (1.18)].

For each $\lambda \geq 0$ we now define an integral operator $F_A(\lambda)$ by

$$(5.1) \qquad (F_A(\lambda)u)(\omega) = \frac{\lambda^{1/4}}{4\pi^{3/2}} \int e^{-i\lambda^{1/2}\omega \cdot \mathbf{r}} A(\mathbf{r})u(\mathbf{r})d\mathbf{r}.$$

$F_A(\lambda)$ is regarded as an operator from $\mathbf{H}$ to $\mathbf{P}$. Similarly, we define $F_B(\lambda)$, using B instead of A in (5.1).

LEMMA 5: $F_A(\lambda) \in B_4(\mathbf{H}, \mathbf{P})$. $F_A(\lambda)$ depends on λ continuously in the sense of $\| \ \|_4$ (i.e., $\lambda \to \mu$ implies $\|F_A(\lambda) - F_A(\mu)\|_4 \to 0$), with $F_A(0) = 0$. Similarly for $F_B(\lambda)$.

Proof: The adjoint operator $F_A(\lambda)^*$ has the adjoint kernel:

$$(5.2) \qquad (F_A(\lambda)^*f)(\mathbf{r}) = \frac{\lambda^{1/4}}{4\pi^{3/2}} A(\mathbf{r}) \int e^{i\lambda^{1/2}\omega \cdot \mathbf{r}} f(\omega)d\,\omega.$$

Hence $F_A(\lambda)^*F_A(\lambda)$ is an integral operator with the composed kernel

$$(5.3) \qquad \frac{1}{16\pi^3} \lambda^{1/2}A(\mathbf{r}) \left[\int e^{i\lambda^{1/2}\omega \cdot (\mathbf{r}-\mathbf{r}')} d\omega \right] A(\mathbf{r}')$$

$$= A(\mathbf{r})A(\mathbf{r}') \sin (\lambda^{1/2}|\mathbf{r} - \mathbf{r}'|)/4\pi^2|\mathbf{r} - \mathbf{r}'|.$$

This is a kernel of the Hilbert-Schmidt type, as is seen by the same argument used in Sec. 2. In fact, we have $\|F_A(\lambda)^*F_A(\lambda)\|_2 \leq M'/\pi$. Hence $F_A(\lambda) \in B_4(\mathbf{H}, \mathbf{P})$ by the definition of B_4, with

$$(5.4) \qquad \|F_A(\lambda)\|_4 \leq (M'/\pi)^{1/2}$$

by (4.4).

To prove the continuity of $F_A(\lambda)$ in λ, we compute (here we omit the subscript A for simplicity)

$$(5.5) \qquad \|F(\lambda) - F(\mu)\|_4^4$$

$$= \|[F(\lambda)^* - F(\mu)^*][F(\lambda) - F(\mu)]\|_2^2$$

$$= \|F(\lambda)^*F(\lambda) + F(\mu)^*F(\mu) - F(\lambda)^*F(\mu) - F(\mu)^*F(\lambda)\|_2^2,$$

which is equal to the square integral of the integral kernel for the operator appearing in $\| \ \|_2$. Here the kernel of $F(\lambda)^*F(\lambda)$ is given by (5.3), similarly for $F(\mu)^*F(\mu)$, whereas the kernels of $F(\lambda)^*F(\mu)$ and $F(\mu)^*F(\lambda)$ are given by

$$(5.6) \quad (\lambda\mu)^{1/4}A(\mathbf{r})A(\mathbf{r}') \sin (|\lambda^{1/2}\mathbf{r} - \mu^{1/2}\mathbf{r}'|)/4\pi^2|\lambda^{1/2}\mathbf{r} - \mu^{1/2}\mathbf{r}'|$$

and its adjoint, respectively, as one verifies easily by a computation similar to (5.3). Thus it is clear that the kernel in question, which is the signed sum of these four kernels, tends to zero pointwise as $\lambda \to \mu$. In view of the fact that $A \in L^3 \cap L^\infty$, it is not difficult to show that the square integral of the sum converges to zero. (We omit the details but note that one would need Vitali's convergence theorem[†] as one of the tools for proof.)

LEMMA 6: *If $u \leftrightarrow \{u(\lambda)\}$ is the spectral representation for $u \in \mathbf{H}$ such that $u(\lambda)$ is a smooth function,*

$$(5.7) \qquad\qquad AE_0'(\lambda)u = F_A(\lambda)^*u(\lambda), \qquad \lambda \geq 0.$$

Proof: The two members of (5.7) are elements of $\mathbf{H}$; note that $u(\lambda) \in \mathbf{P}$ and $F_A(\lambda)^* \in B_4(\mathbf{P}, \mathbf{H})$.

By definition (see the Remark at the end of Sec. 2), we have

$$(5.8) \quad AE_0'(\lambda)u = \lim_{\epsilon\to 0} (2i\pi)^{-1}A[R_0(\lambda + i\epsilon)u - R_0(\lambda - i\epsilon)u].$$

If $u \leftrightarrow \{u(\lambda')\}$, we have

$$R_0(\lambda + i\epsilon)u - R_0(\lambda - i\epsilon)u$$
$$\leftrightarrow \{(\lambda' - \lambda - i\epsilon)^{-1}u(\lambda') - (\lambda' - \lambda + i\epsilon)^{-1}u(\lambda')\}$$
$$= \{2i\epsilon[(\lambda' - \lambda)^2 + \epsilon^2]^{-1}u(\lambda')\},$$

which tends to $\{2i\pi\delta(\lambda' - \lambda)u(\lambda')\}$ as $\epsilon \to 0$. It follows from (5.8) by the Fourier transformation [see (1.16)] that

$$(5.9) \qquad (AE_0'(\lambda)u)(r) = \frac{\lambda^{1/4}}{4\pi^{3/2}}\, A(\mathbf{r}) \int e^{i\lambda^{1/2}\boldsymbol{\omega}\cdot\mathbf{r}}u(\lambda;\omega)\, d\omega$$

$$= (F_A(\lambda)^*u(\lambda)(\mathbf{r}).$$

This proves (5.7).

We are now in a position to give the formula for the S-matrix.

---

† See [5], p. 150.

THEOREM 3: *The S-matrix $S(\lambda)$ is given by*†

(5.10) $\quad S(\lambda) - 1 = -2i\pi\kappa F_A(\lambda)[1 + \kappa Q_0(\lambda + i0)]^{-1}F_B(\lambda)^*.$

$S(\lambda) - 1$ *belongs to the Hilbert-Schmidt class* $B_2(\mathbf{P})$, *and depends on* λ *continuously in* $||\ ||_2$-*norm, with* $S(0) = 1$. $S(\lambda)$ *is unitary for each* λ.

Proof: First we note that the right member of (5.10) belongs to $B_2(\mathbf{P})$ and depends on λ continuously, for $F_B(\lambda)^* \in B_4(\mathbf{P}, \mathbf{H})$ by Lemma 5, $[1 + \kappa Q_0(\lambda + i0)]^{-1} \in B(\mathbf{H})$ by Lemma 1 and (2.18), and $F_A(\lambda) \in B_4(\mathbf{H}, \mathbf{P})$ by Lemma 5, and all these operators depend on λ continuously in their respective norms (see Sec. 4). It is equal to 0 for $\lambda = 0$ since $F_A(0) = 0$ (Lemma 5).

It remains to prove (5.10). Let u, $v \in \mathbf{H}$ be such that their spectral representations $\{u(\lambda)\}$, $\{v(\lambda)\}$ are smooth functions of λ. Then it follows from (3.5) and Lemma 6 that

(5.11) $\quad ((S - 1)u, v)$

$$= -2i\pi\kappa \int_0^\infty ([1 + \kappa Q_0(\lambda + i0)]^{-1}F_B(\lambda)^*u(\lambda), F_A(\lambda)^*v(\lambda))\ d\lambda$$

$$= -2i\pi\kappa \int_0^\infty (F_A(\lambda)[1 + \kappa Q_0(\lambda + i0)]^{-1}F_B(\lambda)^*u(\lambda), v(\lambda))\ d\lambda;$$

note that the inner products (,) under the last integral sign are taken in $\mathbf{P}$. Since the v with smooth $\{v(\lambda)\}$ are dense in $\mathbf{H}$, we must have

$$(S - 1)u \leftrightarrow \{-2i\pi\kappa F_A(\lambda)[1 + \kappa Q_0(\lambda + i0)]^{-1}F_B(\lambda)^*u(\lambda)\}.$$

Since such u form a dense set in $\mathbf{H}$ and since the right member of (5.10) is a smooth function of λ, this proves that S is diagonal in this representation and (5.10) is true.

Remarks: 1. Since $S(\lambda) - 1 \in B_2(\mathbf{P})$, it is an integral operator with a kernel $s(\lambda; \boldsymbol{\omega}, \boldsymbol{\omega}')$, which is square integrable over $\Omega \times \Omega$. This kernel is the composition of the kernels for the three factors on the right of (5.10). If we develop $[1 + \kappa Q_0(\lambda + i0)]^{-1}$ into a power series in κ, we obtain

---

† For this formula cf. [10, 21, 27].

$$(5.12) \quad s(\lambda; \boldsymbol{\omega}, \boldsymbol{\omega}') = -i(8\pi^2)^{-1}\lambda^{1/2}[\kappa \int e^{i\lambda^{1/2}(\boldsymbol{\omega}' - \boldsymbol{\omega}) \cdot \mathbf{r}} V(\mathbf{r}) \, d\mathbf{r}$$

$$- \kappa^2 \iint e^{-i\lambda^{1/2}\boldsymbol{\omega} \cdot \mathbf{r}} V(\mathbf{r}) g_0(\mathbf{r}, \mathbf{r}'; \lambda + i0) V(\mathbf{r}') e^{i\lambda^{1/2}\boldsymbol{\omega}' \cdot \mathbf{r}'} \, d\mathbf{r} \, d\mathbf{r}'$$

$$+ \cdots] \qquad \text{(Born expansion)}$$

in which the factorization $V = AB$ does not appear explicitly.

2. Since $S(\lambda)$ is unitary with $S(\lambda) - 1 \in B_2(\mathbf{P})$, $S(\lambda)$ has a complete set of eigenfunctions, with eigenvalues of the form $e^{2i\delta_n(\lambda)}$ [with real $\delta_n(\lambda)$], $n = 1, 2, \cdots$, which have no cluster point other than 1. The $\delta_n(\lambda)$ are called the *phase shifts*. Since $S(\lambda) - 1$ is continuous in λ in $\| \ \|_2$, the $\delta_n(\lambda)$ are continuous in λ if numbered appropriately. Also we have

$$(5.13) \quad \iint |s(\lambda; \boldsymbol{\omega}, \boldsymbol{\omega}')|^2 \, d\boldsymbol{\omega} \, d\boldsymbol{\omega}' = \|S(\lambda) - 1\|_2^2$$

$$= \sum_{n=1}^{\infty} |e^{2i\delta_n(\lambda)} - 1|^2 = 4 \sum_{n=1}^{\infty} \sin^2 \delta_n(\lambda) < \infty.$$

3. If we assume that $V \in L^1(R^3)$, stronger results are obtained. In this case (5.3) shows that

$$\|F_A(\lambda)^* F_A(\lambda)\|_1 = \text{trace } F_A(\lambda)^* F_A(\lambda)$$

$$= (\lambda^{1/2}/4\pi^2) \int A(\mathbf{r})^2 \, d\mathbf{r} = \lambda^{1/2}\|V\|_{L^1}/4\pi^2$$

if we take $A = |V|^{1/2}$; here $\| \ \|_1$ is the *trace norm*.[†] Hence $F_A(\lambda) \in B_2(\mathbf{H}, \mathbf{P})$. Similarly, one can show that $F_A(\lambda)$ depends on λ continuously in B_2-norm. It follows that $S(\lambda) - 1$ belongs to the *trace class* $B_1(\mathbf{P})$ and depends on λ continuously in the trace norm.

6. SCATTERING CROSS SECTIONS

Let us consider the relationship between $S(\lambda)$ and the scattering cross sections.[‡] It is natural to suppose that the kernel $s(\lambda_0; \boldsymbol{\omega}, \boldsymbol{\omega}_0)$ for $S(\lambda_0) - 1$ is proportional to the scattering amplitude and that $|s(\lambda_0; \boldsymbol{\omega}, \boldsymbol{\omega}_0)|^2 \, d\boldsymbol{\omega}$ is proportional to the differential cross section

---

[†] See [5], pp. 1088ff. and [28].

[‡] This section is not altogether rigorous. It is partly due to the lack of a precise definition of the scattering cross sections.

for scattering into the direction $\boldsymbol{\omega}$ of the incident particle with momentum $\lambda_0^{1/2}\boldsymbol{\omega}_0$. To justify this conjecture and obtain the correct factor, one needs a careful analysis.

Suppose the initial state ϕ_- is such that the particle has approximately a definite momentum $\lambda_0^{1/2}\boldsymbol{\omega}_0$. We may assume that

$$(6.1) \qquad \phi_-(\lambda; \boldsymbol{\omega}) = p(\lambda)q(\boldsymbol{\omega}),$$

where $p(\lambda) = 0$ except when $\lambda \in I_0 = $ a small interval in $(0, \infty)$ and $q(\boldsymbol{\omega}) = 0$ except when $\boldsymbol{\omega} \in \Omega_0 = $ a small part of Ω.

We ask for the probability that in the final state $\phi_+ = S\phi_- \leftrightarrow \{S(\lambda)\phi_-(\lambda)\}$ the particle has momentum directed into a part Ω_1 of Ω. If we denote by P_1 the projection operator of $\mathbf{P}$ onto the subspace $L^2(\Omega_1)$, the required probability is

$$(6.2) \quad w = \int ||P_1 S(\lambda)\phi_-(\lambda)||^2 \, d\lambda = \int ||P_1(S(\lambda) - 1)\phi_-(\lambda)||^2 \, d\lambda,$$

provided Ω_1 and Ω_0 do not meet, for then $P_1\phi_-(\lambda) = 0$. Hence

$$(6.3) \qquad w = \int |p(\lambda)|^2 \, d\lambda \int_{\Omega_1} d\boldsymbol{\omega} \left| \int s(\lambda; \boldsymbol{\omega}, \boldsymbol{\omega}')q(\boldsymbol{\omega}') \, d\boldsymbol{\omega}' \right|^2.$$

To obtain the cross section $\sigma(\lambda_0; \Omega_1, \boldsymbol{\omega}_0)$ for scattering into Ω_1 of the incident particle with the *exact* momentum $\lambda_0^{1/2}\boldsymbol{\omega}_0$, we have to take the limit of w as I_0 and Ω_0 shrink to the points λ_0 and $\boldsymbol{\omega}_0$, respectively. In this procedure, however, the *normalization* of ϕ_- is crucial.

If one does this with ϕ_- kept normalized as $||\phi_-|| = 1$, then $\lim w = 0$. For then we may assume $||p|| = ||q|| = 1$, and q would necessarily tend to 0 *weakly* in $\mathbf{P}$. Hence $(S(\lambda) - 1)q \to 0$ strongly because $S(\lambda) - 1$ is compact, and $w \to 0$ follows.

This is physically quite natural. For in such a procedure the wave function ϕ_- in the configuration space spreads out indefinitely. Since the density thus becomes zero, no scattering can take place. To obtain a definite limit for w, one has to go to the limit in such a way that

$$(6.4) \qquad q(\boldsymbol{\omega}) \sim \delta(\boldsymbol{\omega} - \boldsymbol{\omega}_0),$$

for example. Then (6.3) becomes

$$(6.5) \qquad w \sim \int |p(\lambda)|^2 \, d\lambda \int_{\Omega_1} |s(\lambda; \boldsymbol{\omega}, \boldsymbol{\omega}_0)|^2 \, d\boldsymbol{\omega}.$$

This suggests that I_0 must shrink to λ_0 in such a way that $||p|| =$ const. $= 1$, say. Thus one should choose

(6.6) $$p(\lambda) \sim \delta(\lambda - \lambda_0)^{1/2},$$

for example, rather than $p(\lambda) \sim \delta(\lambda - \lambda_0)$. Then the limiting w is

(6.7) $$w \sim \int_{\Omega_1} |s(\lambda_0; \boldsymbol{\omega}, \boldsymbol{\omega}_0)|^2 \, d\boldsymbol{\omega}.$$

On the other hand, one has to know how the limiting ϕ_- behaves in the configuration space. With (6.1) and (6.4), $\phi_-(\mathbf{r})$ is given by (1.16) as

(6.8) $$\phi_-(\mathbf{r}) \sim \frac{1}{4\pi^{3/2}} \int e^{i\lambda^{1/2}\boldsymbol{\omega}_0 \cdot \mathbf{r}} \lambda^{1/4} p(\lambda) \, d\lambda.$$

Since this is a function of $z = \boldsymbol{\omega}_0 \cdot \mathbf{r}$ alone, it is a wave spread indefinitely in the directions perpendicular to z, but it has a finite "size" in the z-direction. The density of the particle per unit *area* is given by

(6.9) $$\int_{-\infty}^{\infty} |\phi_-(\mathbf{r})|^2 \, dz \sim \frac{1}{16\pi^3} \int_{-\infty}^{\infty} dz |\int e^{i\lambda^{1/2}z} \lambda^{1/4} p(\lambda) \, d\lambda|^2$$

$$= \frac{1}{4\pi^3} \int_{-\infty}^{\infty} dz \left| \int e^{isz} s^{3/2} |p(s^2)| \, ds \right|^2$$

$$= \frac{1}{2\pi^2} \int s^3 |p(s^2)|^2 \, ds = \frac{1}{4\pi^2} \int \lambda |p(\lambda)|^2 \, d\lambda \sim \lambda_0/4\pi^2,$$

where we used (6.6) and Parseval's formula for the Fourier transform in one variable.

According to the usual definition, the cross section $\sigma(\lambda_0; \Omega_1, \boldsymbol{\omega}_0)$ stated above is given by the ratio of (6.7) to (6.9). Thus

(6.10) $$\sigma(\lambda_0; \Omega_1, \boldsymbol{\omega}_0) = 4\pi^2 \lambda_0^{-1} \int_{\Omega_1} |s(\lambda_0; \boldsymbol{\omega}, \boldsymbol{\omega}_0)|^2 \, d\boldsymbol{\omega}.$$

In other words, the differential cross section is given by $4\pi^2 \lambda_0^{-1} |s(\lambda_0; \boldsymbol{\omega}, \boldsymbol{\omega}_0)|^2 \, d\boldsymbol{\omega}$.

The total cross section is conventionally given by setting $\Omega_1 = \Omega$ in (6.10), although this is questionable in view of the assumption made above that Ω_1 should not contain $\boldsymbol{\omega}_0$. The total cross section $\sigma(\lambda_0; \Omega, \boldsymbol{\omega}_0)$, of course, depends on $\boldsymbol{\omega}_0$, the average being equal to

$$(6.11) \quad \frac{1}{4\pi} \int \sigma(\lambda_0; \Omega, \omega_0) \, d\omega_0 = (\pi/\lambda_0) \iint |s(\lambda_0; \omega, \omega_0)|^2 \, d\omega \, d\omega_0$$

$$= (4\pi/\lambda_0) \sum_{n=1}^{\infty} \sin^2 \delta_n(\lambda_0)$$

by (5.13). This is also the value of $\sigma(\lambda_0; \Omega, \boldsymbol{\omega}_0)$ when V is spherically symmetric.

7. POSTSCRIPT

1. The reader will have noticed that we did not consider the (distorted) plane-wave solution of the stationary Schrödinger equation

$$(7.1) \qquad \Delta\phi + (\lambda_0 - \kappa V)\phi = 0, \qquad \lambda_0 > 0.$$

Such functions ϕ are usually defined as the solutions of the integral equation

$$(7.2) \quad \phi(\mathbf{r}) = e^{i\lambda_0^{1/2}\boldsymbol{\omega}_0 \cdot \mathbf{r}} - \kappa \int g_0(\mathbf{r}, \mathbf{r}'; \lambda + i0) V(\mathbf{r}')\phi(\mathbf{r}') \, d\mathbf{r}',$$

in which the first term on the right is the incident plane wave with momentum $\lambda_0^{1/2}\boldsymbol{\omega}_0$ and the second term is the scattered wave. The scattering cross sections are then deduced by using the asymptotic form

$$(7.3) \qquad\qquad |\mathbf{r}|^{-1} f(\lambda_0; \boldsymbol{\omega}, \boldsymbol{\omega}_0) e^{i\lambda_0^{1/2}|\mathbf{r}|}$$

as $|\mathbf{r}| \to \infty$ of the scattered wave, where $\boldsymbol{\omega} = \mathbf{r}/|\mathbf{r}|$.

We did not follow this procedure for several reasons. First, such solutions ϕ are outside of **H** and have no direct physical meaning. Second, it is not easy to prove the existence of the solutions of the integral equations (7.2), at least under the mild assumptions that we made on V, and it is even more difficult to show that the scattered wave has an asymptotic form (7.3). We have been more interested in demonstrating that scattering theory can be satisfactorily worked out within the framework of the Hilbert space **H**.

On the other hand, the existence of the solutions of (7.2) has been proved under slightly stronger assumptions on V, and the relationship between the ϕ and the operators $W_\pm$, S has been established. In particular, it is known that

$$(7.4) \qquad s(\lambda_0; \boldsymbol{\omega}, \boldsymbol{\omega}_0) = i\lambda_0 f(\lambda_0; \boldsymbol{\omega}, \boldsymbol{\omega}_0)/2\pi.$$

For these results see [1, 4, 11, 12, 24, 29].

2. Equation (7.1) is sometimes called the *reduced wave equation*, for it is obtained from the wave equation

$$(7.5) \qquad \frac{\partial^2 \psi}{\partial t^2} = \Delta\psi - \kappa V\psi = -H_\kappa \psi$$

by separating the time t according to $\psi = e^{\pm i\lambda_0^{1/2}t}\phi$. Thus it is not surprising that the solutions ϕ of (7.2) appear also in the scattering theory for the wave equation (7.5). But (7.5) is quite different from the time-dependent Schrödinger equation (1.1). Thus the question arises of how the time-dependent scattering theory for the wave equation is related to that for the Schrödinger equation. The link is contained in the *invariance principle*, which asserts (in this special case) that

$$(7.6) \qquad \operatorname*{s-lim}_{t \to \pm\infty} e^{itH_\kappa^{1/2}}e^{-itH_0^{1/2}} = \operatorname*{s-lim}_{t \to \pm\infty} e^{itH_\kappa}e^{-itH_0} = W_\pm.$$

For details see [2, 14, 16]. For the scattering theory for the wave equation in general, see [22, 31].

3. We considered above only the scattering of a single particle moving in a given potential field. More complicated problems for *multichannel scattering* have been studied by [6, 9, 13].

4. Some of the results of scattering theory can be extended to operators in Banach spaces; see [23].

BIBLIOGRAPHY

This bibliography is not intended to be complete. Stress is laid on recent papers, in which references to older works may be found. The author is indebted to C. L. Dolph for many valuable suggestions.

1. Alsholm, P., and G. Schmidt, "Spectral and scattering theory for Schrödinger operators," Various Publication Series No. 7, Aarhus University, 1969.

2. Bepol'skii, A. L., and M. Sh. Birman, "Existence of wave operators in scattering theory for a pair of spaces," *Izv. Akad. Nauk SSSR*, **32** (1968), 1126–1175 (in Russian); English translation, *Math. USSR-Izv.*, to appear.

3. Birman, M. Sh., "Scattering problems for differential operators with constant coefficients," *Functional. Anal. i Priložen.*, **3** (1969), 1–16 (in Russian); English translation, *Functional Anal. Appl.*, to appear.

4. Buslaev, V. S., "Trace formulas for the Schrödinger operator in a three dimensional space," *Dokl. Akad. Nauk SSSR*, **143** (1962), 1067–1070 (in Russian); English translation, *Soviet Physics Dokl.*, **7** (1962), 295–297. See also *Topics in Math. Physics*, **1** (1967), 69–85.

5. Dunford, N., and J. T. Schwartz, *Linear Operators*, Parts I and II. New York: Interscience, 1958, 1963.

6. Faddeev, L. D., "Mathematical questions in the quantum theory of scattering for a system of three particles," *Trudy Mat. Inst. Steklov.*, **69** (1963), 1–122. Translated from the Russian by Ch. Gutfreund. Translation edited by L. Meroz. Israel Program for Scientific Translations, Jerusalem; Daniel Davey and Company, Inc., New York, 1965.

7. Friedrichs, K. O., "Perturbation of spectra in Hilbert space," *Amer. Math. Soc. Lectures in Appl. Math.*, Vol. III, Providence, R.I., 1965.

8. Helson, H., *Lectures on Invariant Subspaces*. New York: Academic Press, 1964.

9. Hepp, K., "On the quantum mechanical N-body problem," *Helv. Phys. Acta*, **42** (1969), 425–458.

10. Howland, J. S., "A perturbation-theoretic approach to eigenfunction expansions," *J. Functional Anal.*, **2** (1968), 1–23.

11. Ikebe, T., "Eigenfunction expansions associated with the Schrödinger operators and their applications to scattering theory," *Arch. Rational Mech. Anal.*, **5** (1960), 1–34.

12. ———, "On the phase-shift formula for the scattering operator," *Pacific J. Math.*, **15** (1965), 511–523.

13. Jauch, J. M., "Theory of the scattering operator," *Helv. Phys. Acta*, **31** (1958), 127–158; 661–684.

14. Kato, T., *Perturbation Theory for Linear Operators*. Berlin-Heidelberg-New York: Springer, 1966.

15. ———, "Wave operators and similarity for some non-selfadjoint operators," *Math. Ann.*, **162** (1966), 258–279.

16. ———, "Scattering theory with two Hilbert spaces," *J. Functional Anal.*, **1** (1967), 342–369.

17. ———, "Some results on potential scattering," *Proc. International Conference on Functional Analysis and Related Topics*, Tokyo, 1969.

18. Kato, T., and S. T. Kuroda, "Theory of simple scattering and eigenfunction expansions," in *Functional Analysis and Related Fields*. New York: Springer, 1970, pp. 99–131.

19. Kuroda, S. T., "On the existence and the unitary property of the scattering operator," *Nuovo Cimento*, **12** (1959), 431–454.

20. ———, "An abstract stationary approach to perturbation of continuous spectra and scattering theory," *J. D'Analyse Math.*, **20** (1967), 57–117.

21. ———, "Some remarks on scattering for Schrödinger operators," *J. Fac. Sci. Univ. Tokyo*, Sec. IA, **17** (1970), 315–329.

22. Lax, P. D., and R. S. Phillips, *Scattering Theory*. New York: Academic Press, 1967.

23. Lin, S. C., "Wave operators and similarity for generators of semigroups in Banach spaces," *Trans. Amer. Math. Soc.*, **139** (1969), 469–494.

24. Povzner, A. Ya., "On expansions in functions which are solutions of a scattering problem," *Dokl. Akad. Nauk SSSR*, **104** (1955), 360–363 (in Russian).

25. Putnam, C. R., *Commutation Properties of Hilbert Space Operators and Related Topics*. New York: Springer, 1967.

26. Rejto, P. A., "On partly gentle perturbations III," *J. Math. Anal. Appl.*, **27** (1969), 21–67.

27. Scadron, M., S. Weinberg, and J. Wright, "Functional analysis and scattering theory," *Phys. Rev.*, **135** (1964), B202–207.

28. Schatten, R., *Norm Ideals of Completely Continuous Operators*. Berlin-Göttingen-Heidelberg: Springer, 1960.

29. Simon, B., "Quantum mechanics for Hamiltonians defined as quadratic forms," Dissertation, Princeton University, 1970.

30. Sobolev, S. L., "Applications of functional analysis in mathematical physics," *Transl. Math. Monogr.*, Vol. 7. Providence, R.I.: Amer. Math. Soc., 1963.

31. Wilcox, C. H., "Wave operators and asymptotic solutions of wave propagation problems of classical physics," *Arch. Rational Mech. Anal.*, **22** (1966), 37–78.

32. ———, "Perturbation theory and its applications in quantum mechanics," Publ. No. 16, Math. Research Center, University of Wisconsin. New York-London-Sydney: Wiley, 1966.

DYNAMICS OF SELF-GRAVITATING SYSTEMS: STRUCTURE OF GALAXIES†

C. C. Lin‡

1. INTRODUCTION

In ancient times, the study of the behavior of celestial objects perhaps provided one of the principal motivations for the study of mathematics. Through the ages there has been a continual interplay between mathematics and astronomy. Newton's study of planetary motion, which led him from Kepler's laws of planetary motion, through exact mathematical deductions, to the law of universal gravitation, was made concurrently with his development of the infinitesimal calculus. The solar system is indeed only an atom in the universe; yet the law of universal gravitation, discovered within it, appears to be valid for much larger scales, until we reach the scale of the whole universe. But here mathematics

† The eighth John von Neumann Lecture delivered at the 1967 SIAM Summer Meeting on August 30, 1967, in Toronto, Ontario.

‡ Department of Mathematics, Massachusetts Institute of Technology, Cambridge, Massachusetts 02139. The writing of this paper was supported in part by a grant from the National Science Foundation.

again plays an essential role. Riemannian geometry provided the framework for the formulation of Einsteinian cosmology.

This evening, I shall discuss the mathematical studies of a class of astronomical phenomena on distance scales much larger than that of the solar system, but much smaller than that of the universe as a whole, so that Newton's law of universal gravitation still holds. I am going to discuss the mathematical theory of systems of stars, either in the form of globular *clusters*, consisting of only hundreds of thousands of stars, or in the form of galaxies, consisting of up to hundreds of billions of stars.

Globular clusters are spherical objects, relatively small by our standards, being only about 10^2–10^3 light years in diameter. There are over a hundred of them in our own galaxy, which has a diameter of the order of 10^5 light years. Galaxies are in clusters, and they are separated from each other by distances of the order of ten galactic diameters. In the observable part of the universe, there are estimated to be three billion galaxies. (See Allen [1, p. 276].)

A system of stars separated by distances of the order of 10^6–10^7 of their radii may be treated as a system of point masses with mutual gravitational attraction. Thus, we have the classical N-body problem for globular clusters and for galaxies. According to the laws of Newton, we may write down the following system of equations governing their motion:

$$(1.1) \qquad \frac{d^2\mathbf{x}_\alpha}{dt^2} = G \sum_{\beta \neq \alpha} \frac{\partial}{\partial \mathbf{x}_\alpha} \frac{m_\beta}{|\mathbf{x}_\alpha - \mathbf{x}_\beta|}.$$

In the above system of equations, m_β is the mass of the βth star; $\mathbf{x}_\beta(t)$ is its positional vector at time t relative to the inertial ("fixed") coordinate system of the observer; G is the constant of universal gravitation; Σ denotes a summation with respect to the index β over the range 1, 2, $\cdots$, N (with the exception of α), where N is the total number of stars in the system. In (1.1), we have not included the forces due to other external agencies, which should be examined, for example, in the case of star clusters in our galactic system. We note that the mass m_α of the particle in question has been canceled from both sides of (1.1).

The problem is thus clearly to develop a mathematical theory of dynamical systems, all the basic physical laws having been satisfied by stipulating (1.1). Furthermore, it is natural to deal with a system composed of a vast number of stars by using the statistical approach:† It is well-known that such subjects as the three-body problem, integrals of motion, the ergodic hypothesis, Brownian motion, etc., are topics which received attention not only from John von Neumann but also from other mathematicians, notably George David Birkhoff, Norbert Wiener and Henri Poincaré (to mention only the deceased).

But there is a more specific problem that interested John von Neumann. This is the relaxation process of a stellar system, in particular, the diffusion of the distribution function in the velocity space. It can be best explained by considering globular clusters. In a globular cluster, the stars are sufficiently far apart from one another, so that they may be treated as point masses. On the other hand, the stars in a globular cluster are also sufficiently near to each other that close encounters of the stars are sufficiently frequent, energy and momentum are exchanged between the stars, and the system tends toward some kind of statistical equilibrium. In this process of adjustment, or relaxation, it is expected that some stars will be ejected. The evolution of globular clusters via this relaxation process is indeed an impressive and interesting stochastic process provided by Nature. It is this process which interested John von Neumann, who collaborated with Chandrasekhar in the early stages of the development of a mathematical theory (Chandrasekhar and von Neumann [5]). Subsequently, Chandrasekhar [2] carried out the complete development of this theory, which has remained a classic in this subject, and has influenced the development of plasma physics as well. Recently, there have been various attempts to refine Chandrasekhar's theory and to improve upon it by the inclusion of collective effects. So far, there have been only minor modifications introduced, but various investigations

---

† The study of N-body problems, not using the particular statistical approach discussed in this lecture, has also been pursued by many astronomers with the use of modern computer facilities. A great deal of activity is continuing at the present time.

are still going on. We shall return to mention some of them in the proper contexts in the discussions in Secs. 6 and 7.

For the present, let us consider those stellar systems for which the relaxation process is not important: the galaxies such as the Milky Way system that we live in. We wish to discuss in particular the following two problems:

(*i*) What is the nature of the spiral pattern observed in disk-shaped galaxies?

(*ii*) Why are some of the galaxies in the shape of a rotating bar? In studying the first problem, we find ourselves dealing with the concept of *collective modes*, which are important in plasma physics. In the study of the second problem, we find that insight may be gained from the classical studies of rotating fluid masses done by Jacobi, Riemann and Dedekind.†

1.1. FORMULATION OF A MATHEMATICAL THEORY: COLLISIONLESS APPROXIMATION. In order to discuss the issues just described in more concrete terms, let us consider in some detail the general outline of a statistical description of stellar systems. As already mentioned, in many cases, especially in galaxies, it is entirely justifiable‡ to neglect the effect of close encounters altogether over a period of 10^{10} years (which is the time scale of the universe). We then have only to consider the smooth gravitational potential $V(\mathbf{x}, t)$ of the stellar system. Each star moves according to the following system of differential equations:

$$(1.2) \qquad\qquad \frac{d\mathbf{x}}{dt} = \mathbf{v},$$

$$(1.3) \qquad\qquad \frac{d\mathbf{v}}{dt} = \mathbf{g},$$

where

$$(1.4) \qquad\qquad \mathbf{g} = -\nabla V.$$

---

† For a review of the current developments along these lines, see Chandrasekhar [3], [4] and Lebovitz [20]. The first reference also gives a very interesting account of the historical developments.

‡ A simple argument was given by Lin [22]. For a detailed analysis, see Chandrasekhar [2]. Cf. Secs. 6 and 8.

These equations may be regarded as the smoothed-out form of (1.1).

For a statistical treatment, we introduce a distribution function $\Psi(\mathbf{x}, \mathbf{v}, t)$, which is the number of stars per unit volume of the phase space. We further assume, for simplicity, that all the stars have the same mass m_*. Then the mass density in the configurational space is given by

$$(1.5) \qquad \rho_*(\mathbf{x}, t) = m_* \int \Psi \, d\tau(\mathbf{v}).$$

It is easy to see that Ψ satisfies the collisionless Boltzmann equation

$$(1.6) \qquad \frac{\partial \Psi}{\partial t} + (\mathbf{v} \cdot \nabla)\Psi + (\mathbf{g} \cdot \nabla_v)\Psi = 0,$$

where $\mathbf{g}$ is given by (1.4), while V is determined by the Poisson equation

$$(1.7) \qquad \nabla^2 V = 4\pi G \rho_* = 4\pi G m_* \int \Psi \, d\tau(\mathbf{v}).$$

Clearly, the system of equations (1.4), (1.6), and (1.7) is nonlinear in Ψ. But if $\mathbf{g}$ is regarded as known in (1.6), the equation is formally linear in Ψ, and its characteristics are given by (1.2) and (1.3) for the motion of a single particle. The system may thus be said to be quasi-linear in Ψ.

The theory of collision-free stellar systems has wide practical applicability and will be adopted in most of the subsequent discussions. There are many interesting general features such as collective modes and Landau damping. For a simple example, see Appendix A.

2. NORMAL SPIRALS: QSSS HYPOTHESIS

The majority of galaxies are disk-shaped, and most of them exhibit two-armed spiral structure over the whole disk. These are termed normal spirals in the Hubble classification. Since I have discussed these problems with mathematical audiences during the past few years, in particular with a SIAM meeting in New York (Lin [22]), and with the summer school at Ithaca

(Lin [23]), I shall only briefly sketch the problem here in order to discuss some recent progress.†

We consider a cylindrical coordinate system $(\check{\omega}, \theta, z)$ with the plane $z = 0$ coinciding with the galactic disk. Let $c_{\check{\omega}}$ and c_θ be the peculiar velocity of the stars relative to a differentially rotating disk with angular velocity $\Omega(\check{\omega})$. We then consider a small disturbance from the symmetrical distribution $\Psi_0(\check{\omega}, c_{\check{\omega}}, c_\theta)$ so that the distribution function may be written in the form

$$(2.1) \qquad \Psi = \Psi_0(1 + \psi), \qquad \psi \ll 1.$$

The basic distribution function Ψ_0 is taken to be given; its determination will be discussed in Sec. 4.

We attempt to find self-sustained modes with ψ of the form

$$(2.2) \qquad \psi = \hat{\psi}(\varpi, c_{\check{\omega}}, c_\theta)e^{i(\omega t - m\theta)}.$$

The corresponding gravitational potential in the plane $z = 0$ must be of the form $V = V_0(\varpi) + V_1(\varpi, \theta, t)$, where

$$(2.3) \qquad V_1 = A(\varpi) \exp \{i[\omega t - m\theta + \Phi(\varpi)]\},$$

where m is an integer, and $A(\varpi)$ and $\Phi(\varpi)$ are real.

One important point to be noted is that, if $\Phi(\varpi)$ *varies rapidly in ϖ in a suitable manner, tightly wound spirals are naturally obtained* with m as the number of arms. Another point that will later turn out to be important is that the energy and angular momentum *integrals for the basic state* (represented by Ψ_0) *would remain constants along the characteristics of the linearized differential equation for ψ.*

It has been found convenient to carry out the investigation with the help of the QSSS hypothesis (the hypothesis of quasi-stationary spiral structure). The basic reason for doing this is to make it easier to study the joint influence of gas and stars and to make a proper evaluation of their relative importance. The scheme is shown in Fig. 1.

The central part of the calculation is then the response of a stellar sheet to a spiral gravitational field in the form (2.3).

---

† Some more details were actually presented at the lecture by quoting from the publications cited and from the papers by Lin and Shu [26] through [28].

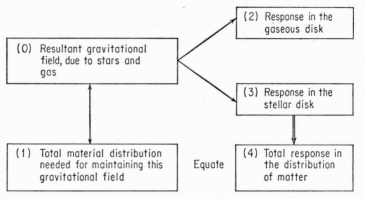

FIG. 1.

Detailed calculations show that, for an ellipsoidal (Schwarzschild) distribution of velocities, the amplitude of the density variation is given by

$$(2.4) \qquad \frac{\acute{\sigma}}{\sigma_*} = -\frac{k^2 A}{\kappa^2 - (\omega - m\Omega)^2} \, \mathfrak{F}_\nu(x),$$

where $k = \Phi'(\varpi)$, κ is the epicyclic frequency defined by

$$(2.5) \qquad \kappa^2 = (2\Omega)^2 \left\{ 1 + \frac{\varpi}{2\Omega} \frac{d\Omega}{d\varpi} \right\},$$

and

$$(2.6) \quad \mathfrak{F}_\nu(x) = \frac{1 - \nu^2}{x} \left\{ 1 - \frac{\nu\pi}{\sin \nu\pi} \cdot \frac{1}{2\pi} \int_{-\pi}^{\pi} \cos (\nu s) e^{-x(1 + \cos s)} \, ds \right\}.$$

In the above formulas,

$$(2.7) \qquad \nu = m(\Omega_p - \Omega)/\kappa, \qquad \Omega_p = \omega/m,$$

$$(2.8) \qquad x = k^2 \langle c_\varpi^2 \rangle / \kappa^2.$$

It was pointed out that $\mathfrak{F}_\nu(x)$ may be regarded as a *reduction factor* which measures the reduction in the effectiveness of the stars to participate in the spiral distribution as its dispersion speed $\langle c_\varpi^2 \rangle$ is increased (cf. Fig. 2).

From this calculation of response, it follows that the *dispersion relationship* is given by

(2.9)
$$\frac{|k|}{k_*} = \frac{1 - \nu^2}{(\sigma_0/\sigma_*) + \mathfrak{F}_\nu(x)}$$

with

(2.10)
$$k_* = \kappa^2/(2\pi G \sigma_*),$$

where σ_0 is the gaseous density while σ_* is the stellar density. The effect of turbulence is neglected in (2.9). If it were included, there would be a similar reduction factor for the gas. It can be shown that the case $\sigma_0/\sigma_* = 0$ is quite typical for all the discussions of the implications of the dispersion relationship.

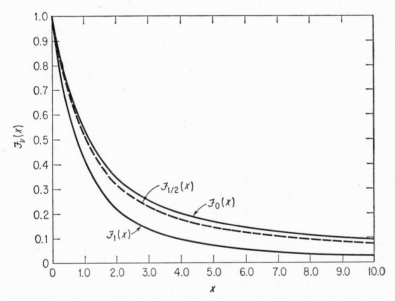

FIG. 2. The reduction factor. The abscissa is a measure of the mean square value of the radial component of the peculiar velocity of the stars.

The dispersion relationship (2.9) implies also the results for instability ($\nu^2 < 0$). At the margin of stability ($\nu^2 = 0$), there is a minimum value of $\langle c_\varpi^2 \rangle$ required for stability to hold at all wavelengths (see Toombe [45]). If this minimum value of $\langle c_\varpi^2 \rangle$ is chosen, then the above dispersion relationship becomes one involving only k/k_* and $|\nu|$. This is calculated by Frank Shu (Lin and Shu [28]) and plotted in Fig. 3.

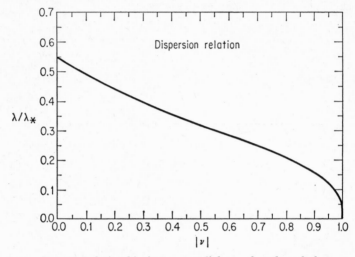

FIG. 3. Relationship between radial wavelength and the angular frequency at which the stars encounter the pattern ($\lambda\ 2\pi/k, \lambda_* + 2\pi/k_*$ as defined in Sec. 2).

The theory thus yields a very definite relationship, except for a single numerical parameter Ω_p. Once its value is chosen, we have a *complete pattern over the whole galactic disk.* Thus, there is a stringent requirement to be fulfilled. (One may say it is an infinity of numerical requirements to be satisfied by the choice of one numerical value.) An example of such a pattern is shown in Fig. 4 for the Milky Way system. Detailed comparisons with observations have been made and will be briefly discussed in the next section. At this point, let us refer briefly to a few mathematical points.

1. It is worth noting that the method adopted here automatically leads to the treatment of *collective modes;* i.e., the statistical distribution of the stars into a spiral configuration

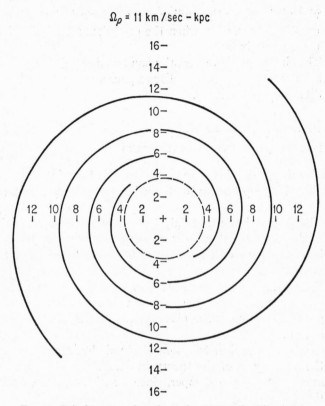

FIG. 4. Spiral pattern based on the 1965 Schmidt model for our own galaxy.

supplies the spiral gravitational field, which also acts as the agent for producing the spiral distribution of the stars. This should be contrasted with the approach adopted by B. Lindblad [30] and that of others who studied numerically the motion of a system of stars.

2. The theory is still limited to its linearized form. Even in this form, some of the nonlinear nature of the problem is included, but a full nonlinear theory is as yet not available.

3. Within the linear theory, the solution of the problem of the stellar response is an interesting one from a technical point of

view; it involves the finding of a regular solution of a partial differential equation when the coefficients have singularities (see Appendix B).

4. The theory of characteristics can be used with advantage in solving the problem of stellar response, as shown recently by Shu (see Appendix C).

3. THE MILKY WAY SYSTEM: MIGRATION OF INDIVIDUAL STARS

Application of the theory to the observed pattern in our galaxy—the Milky Way—produces the following satisfactory results:

1. The theory shows that only two-armed patterns can exist to any noticeable extent. (This is true not only for the Galaxy, but for others with mass distributions that are similar in a loose sense.)
2. By taking a pattern speed equal to (about) one-half the angular speed in the solar vicinity, the pattern obtained is in good agreement with observations throughout the whole galaxy.
3. The motion and distribution of atomic hydrogen agree with the predictions of the theory in detail. (This is outside the scope of the present paper, since we wish to concentrate on the stars.)
4. The migration of individual stars to their places of formation leads to reasonable results only if our theory is adopted.

This last point is a very intersting one from the general point of view. By and large, we have studied the statistical distribution of stars, treated similarly to molecules in a gas. It is unusual to study the behavior of *individual* molecules. However, since we on the earth are of "subatomic" dimensions in this case, we are indeed interested in the study of the individual behavior of such molecules. In this case, we wish to trace the stars backwards in time to their places of formation.

We know that stars are formed at places where the gas concentration is high. We also know that the stars have been moving largely under the influence of the collective gravitational field. Thus, if we know this field and if we know the precise age of the stars, we can determine where a star, now in our neighborhood, was formed, provided we know its present location and its present velocity.

The Danish astronomer Bengt Strömgren recently developed a vastly improved spectral method for the accurate determination of stellar ages. He and his collaborators were therefore able to carry out the above-mentioned program. However, in the absence of any idea of the spiral structure, they assumed that the stars move in the mean gravitational field. This led to unreasonable results. We have repeated the analysis with a spiral field included, at a strength equal to a few per cent of the symmetric field. *All* the stars observed are found to lie in spiral arms, where they are expected.

For further details, the reader is referred to the paper by Lin, Shu, and Yuan [29]. The work was reported at a meeting of the International Astronomical Union in Prague a week prior to this lecture.

4. CONSTRUCTION OF EQUILIBRIUM MODELS: CASES WITH AXIAL SYMMETRY

When close encounters of stars can be neglected, stellar systems in equilibrium can be described by the time-independent form of (1.6):

$$(4.1) \qquad (\mathbf{v}\cdot\nabla)\Psi + (\mathbf{g}\cdot\nabla_v)\Psi = 0.$$

Jeans' theorem then states that every solution of (4.1) is of the form

$$(4.2) \qquad \Psi = F(I_1, I_2, \cdots),$$

where the I's are the isolating integrals† of motion for a single particle. The one-particle energy

$$(4.3) \qquad E = \tfrac{1}{2}\mathbf{v}^2 + V$$

_____

† Isolating integrals are single-valued integrals that correspond to zero extension in the six-dimensional phase space.

is such an integral, as are the components of angular momentum along any axis of symmetry that the system might have. The classical problem in equilibrium stellar dynamics, as formulated by Jeans, is to determine a potential $V(\mathbf{r})$ that is consistent with the Poisson's equation and a suitable form of the function F.

Much work has been done on the problem of the existence of integrals of motion besides the integrals of energy and angular momentum. Moser and Arnold (see Contopoulos [7]) proved certain general theorems, under rather stringent requirements. Contopoulos [7] proved theorems involving asymptotic approximation. He proved that to each Hamiltonian for the mechanical system, there always exists another Hamiltonian, which is arbitrarily close to the given one, and for which the third integral exists. He and his collaborators also made extensive numerical calculations to observe the changeover from existence to non-existence of the third integral as the parameters of the problem are changed gradually. It emerged from his investigations that one has to introduce the concept of quasi-isolating integrals in order to describe all the cases adequately.

4.1. SPECIFIC MODELS. Observationally, we know that there are (a) spherical systems, (b) axially symmetrical systems (often with a superposed spiral pattern) and (c) barred systems. Elliptical galaxies, with various degrees of flattening, are designated E0 (which are spherical) to E7, which is almost a disk. The general mechanism of flattening of stellar systems by rotation was discussed by Poincaré and by Jeans [13]. Jeans also pointed out, by analogy with kinetic theory of gases, that in the state of statistical equilibrium of a stellar system, the stellar velocities should be in an *isotropic Gaussian distribution relative to a state of uniform mean rotation*. Indeed, this is the "most probable" configuration in the usual sense of statistical mechanics. In both cases, cylindrical symmetry is assumed. Recently, Milder [38] discussed an extension of this principle to configurations without axial symmetry. To what extent such theorems would be relevant to galaxies is a serious question. We know that galaxies are giant stellar systems in which stellar collisions are so infrequent that the equilibrium configuration may never be reached. But they are certainly relevant to globular clusters (Sec. 7).

Let us now discuss these special cases in a little more detail.

Case (a). *Spherical symmetry.* In the case of spherical symmetry, the integrals of motion are the energy and all the three components of angular momentum. However, if Ψ is to have spherical symmetry, the latter can only enter through an invariant quantity, such as the square of the magnitude of the angular momentum. Thus,

$$(4.4) \qquad \Psi = F(E, J),$$

where J is the magnitude of the angular momentum.

Examples of solutions with spherical symmetry have been studied extensively by King [17], Lynden-Bell [31 through 35] and many others.

Case (b). *Axial symmetry.* If the system has considerable angular momentum, the distribution in configurational space would be expected to have only axial symmetry. Then the obvious integrals of motion for a single particle are only the energy and the component of angular momentum parallel to the axis of symmetry. Thus, a class of solutions is

$$(4.5) \qquad \Psi = F(E, J_z),$$

where J_z is the z-component of the single-particle angular momentum; we have, of course, taken the z-axis parallel to the axis of symmetry. Models of elliptical galaxies have recently been constructed by Prendergast and by Ng [39] of Columbia University along these lines.

Case (c). *Disks.* When the angular momentum is very large, the ellipsoidal shape tends to degenerate into a disk. Now, when the thickness is very much smaller than the scales in the plane of the disk, the motion in the z-direction is essentially that of a *one-dimensional* sheet. It is to be expected then that there should be integrals of motion in the z-direction. Thus, we have a *third* integral of motion besides the energy and angular momentum integrals discussed before.

As the thickness increases, the third integral would be expected to persist in a modified form when the thickness is not large; but there is no reason to expect that it will persist forever. This is an interesting problem by itself; it is also important from a practical point of view when one wants to construct a complete model for disk-shaped galaxies, like our own. Indeed, this is the problem that motivated Contopoulos to the studies mentioned above.

Another natural approach to the construction of disk models is to start with the infinitesimally thin case, and then proceed to the case of a small but finite thickness by asymptotic approximation. This is the method adopted by Vandervoort [46], who showed how disk models can be constructed for a fairly general class of rotation curves. The preliminary step is to construct a model disk of zero thickness and to assume that the stars have no dispersion velocity. Various methods have been developed for this problem over a number of years. Some authors use the spheroidal coordinate system, others the cylindrical coordinates. As could be easily expected, the former approach tends to produce a model with density distributions more or less confined to a finite disk and tapering off at infinity rather suddenly, while the latter tends to produce a model with a more gradual dropping of density. In either case, because of the long-range nature of the forces, the observed velocity distribution ("rotation curve") must be known for the whole disk before the density distribution can be evaluated with confidence. For further detailed guide to references, see a review article by Lin [24].

The best model of an infinitesimally thin disk has been constructed by Frank Shu [41]. Given any model with zero dispersion velocity, his method yields an exact model in terms of the isolating integrals such that the local distribution at any point is close to the Schwarzschild distribution (which is known to be approximately observed). The amount of dispersion velocity can be specified as any reasonable function of the radial distance.

5. ROTATING SYSTEMS WITH AXIAL SYMMETRY:
   BAR-SHAPED GALAXIES

5.1. ROTATING STELLAR SYSTEMS AND ROTATING FLUID MASSES. As might be expected from general considerations, the galactic systems are actually rotating, except perhaps the spherical galaxies.[†] It is easy to conceive of the galaxies as approximately represented by a family of rotationally symmetric systems, with increasing degree of flatness as its angular momentum is increased. The barred spirals came as somewhat of a surprise, since the system of stars is *not* symmetrical about the axis of rotation, and yet the bar apparently holds together as a permanent entity.

However, one would not be so surprised if one were familiar with the work of the mathematicians of the nineteenth century: Jacobi, Riemann, and Dedekind. These people did not even know of the existence of galaxies when they did their work; the concept of island universes of the philosopher Kant was only a speculation at that time. They were interested in the configuration and stability problems associated with stars and planets, and hence were concerned with self-gravitating rotating masses of *liquid* with isotropic pressure. It was known to Maclaurin that a homogeneous liquid spheroid, rotating about its axis of symmetry, can be a configuration of equilibrium. It was first shown by Jacobi that, when the angular velocity is sufficiently high, a rotating ellipsoid with three unequal axes is also a possible form of equilibrium. Indeed, there is a bifurcation of solutions branching off from a common configuration belonging to both families of ellipsoids. At high speeds of rotation, the Jacobi ellipsoids are like bars, rotating about a short axis.

Dedekind and Riemann found other rotating ellipsoids. Dedekind also proved a theorem on the existence of "adjoint" fluid motions for the same ellipsoidal figure, corresponding to the transpose of a certain matrix (see Chandrasekhar [3]). The work of Riemann was extended further by Chandrasekhar [3, 4] and by Lebovitz [19]. Their work also serves to clarify the general outlook of the whole subject. For other recent work concerning the equilibrium and stability of rotating fluid masses, both compressible and incompressible, the reader is referred to a review article by Norman Lebovitz [20] for the details and for the references.

---

† Even here, the inference is not necessarily correct (cf. Lynden-Bell [31]: "Can spherical clusters rotate?").

We wish only to stress one point. The existence of bifurcation of solutions in the case of rotating masses of *liquid* suggests the existence of a similar phenomenon for a gas, and possibly for a stellar system. However, this is not necessarily the case. Indeed, Jeans [13] estimated that for polytropic gas, bifurcation can occur only if the polytropic index *exceeds* 2.2. This value has been very closely verified recently by James [12], by a much more elaborate analysis. We have thus to examine stellar systems for the possibility of bifurcation of equilibrium solutions, and for solutions with bar shapes. Some recent progress in this work has been accomplished by K. C. Freeman [9], but the problem has not been completely solved.

5.2. BARRED GALAXIES: RECENT DEVELOPMENTS.

The first problem in describing barred spirals is to show that a rotating stellar system can maintain equilibrium in a barred configuration. Once this is established, the gas streaming is not difficult to describe (Prendergast [40], Freeman [8], Freeman and Mestel [10]).

The dynamical theory of the structure and evolution of barred spiral galaxies was investigated by Freeman [9] in a series of three papers. A different model is considered in each of the three papers, but all are self-gravitating, rotating, and collisionless stellar systems. In Paper I, the configuration of an elliptical cylinder is considered; in Paper II, an ellipsoid; in both cases, the density distribution is homogeneous. In Paper III, a family of inhomogeneous elliptical disks is described. All three models lead to the same conclusion: if the arms of barred spiral galaxies form from gas streaming out along the bar, the bar will become more dense and more stubby, and will rotate more rapidly, as it loses mass, energy and angular momentum.

At first sight, the ellipsoidal configuration in Freeman's Paper II (see [9]) is the best. Unfortunately, he has to impose the restriction that the centrifugal and gravitational forces balance along the major axis. Freeman stated (introduction of Paper III), "We have not been able to construct exact, homogeneous, uniformly rotating ellipsoids with no diagonal component of the velocity dispersion vanishing identically. This would have been the most realistic model, for a bar, which our method of constructing models could have produced."

6. THEORY OF COLLISIONS

Chandrasekhar [2] evaluated the effect of the fluctuating field through a consideration of binary collisions. He was able to show that the cumulative effect can be represented in the form

$$(6.1) \qquad \left(\frac{\partial \Psi}{\partial t}\right)_{\text{coll.}} = \sum_{i=1}^{3} \frac{\partial}{\partial v_i}\left(q \frac{\partial \Psi}{\partial v_i} + \eta \Psi v_i\right),$$

where q is a diffusion coefficient in velocity space, and η is the coefficient of dynamical friction, inversely proportional to the relaxation time. It can be easily verified that the collision term vanishes identically for the Maxwell-Boltzmann distribution

$$(6.2) \qquad \Psi = A \exp(-\beta E),$$

where A is a constant, and $E = \frac{1}{2}\mathbf{v}^2 + V$, provided that $q\beta = \eta$. (All parameters here may, of course, also vary slowly in configurational space.)

Instead of repeating a development of Chandrasekhar's theory here (which is well-known), we shall describe the outline of an approach due to C. S. Wu [47] (who adapted Klimentovitch's discussion of electromagnetic plasmas to the stellar case) to indicate the nature of the problem.

Following the usual practice in statistical mechanics, we consider an ensemble of identical systems and speak of one realization when we refer to one particular system as one member of this ensemble. For each particular realization, we may then write down the equations of motion of each particle in the form

$$(6.3) \qquad \frac{d\mathbf{x}_\alpha}{dt} = \mathbf{v}_\alpha,$$

$$(6.4) \qquad \frac{d\mathbf{v}_\alpha}{dt} = \mathbf{g}_m(\mathbf{x}_\alpha, t),$$

where $\alpha = 1, 2, \cdots, N$, and

$$(6.5) \qquad \mathbf{g}_m(\mathbf{x}, t) = - \sum_\beta Gm_* \frac{(\mathbf{x} - \mathbf{x}_\beta)}{|\mathbf{x} - \mathbf{x}_\beta|^3};$$

here the summation over β extends over all the particles if $\mathbf{x} \neq \mathbf{x}_\alpha$ but with the exclusion of $\beta = \alpha$, if $\mathbf{x} = \mathbf{x}_\beta$.

Let us now define the microscopic number density function in phase space $(\mathbf{x}, \mathbf{v})$:

$$(6.6) \qquad n(\mathbf{x}, \mathbf{v}, t) = \sum_\alpha \delta(\mathbf{x} - \mathbf{x}_\alpha(t))\delta(\mathbf{v} - \mathbf{v}_\alpha(t)),$$

where the summation extends over all the particles, and δ is the Dirac delta function in three dimensions. Then the ensemble average $\langle n \rangle$ of n, the mean number density in phase space, may be identified with the function Ψ defined previously:

$$(6.7) \qquad \langle n \rangle = \Psi(\mathbf{x}, \mathbf{v}, t).$$

In terms of (6.6), we may write down the following formal partial differential equation to replace (6.3) and (6.4):

$$(6.8) \qquad \frac{\partial n}{\partial t} + (\mathbf{v} \cdot \nabla)n + (\mathbf{g}_m \cdot \nabla_v)n = 0,$$

and the following equation in place of (6.5):

$$(6.9) \qquad \nabla \cdot \mathbf{g}_m = -4\pi Gm_* \int n \, d\tau(\mathbf{v}).$$

Let us now write all microscopic quantities, such as n and $\mathbf{g}_m$, as the sum of an *average* and a *fluctuation*:

$$(6.10) \qquad n = \langle n \rangle + \tilde{n}, \qquad \mathbf{g}_m = \langle \mathbf{g} \rangle + \tilde{\mathbf{g}}.$$

Then $\langle \mathbf{g} \rangle$ may be identified with $\mathbf{g}$ defined in (1.4). If we now take the ensemble average of (6.8), we find

$$(6.11) \qquad \frac{\partial \Psi}{\partial t} + (\mathbf{v} \cdot \nabla)\Psi + (\mathbf{g} \cdot \nabla_v)\Psi = -\nabla_v \cdot (\tilde{n}\tilde{\mathbf{g}}),$$

$$(6.12) \qquad \nabla \cdot \mathbf{g} = -\nabla^2 V = -4\pi Gm_* \int \Psi \, d\tau(\mathbf{v}),$$

where (6.12) reproduces (1.7). The equations governing the fluctuating quantities may then be obtained by subtracting (6.11) and (6.12) from (6.8) and (6.9). We obtain

$$(6.13) \quad \frac{\partial \tilde{n}}{\partial t} + (\mathbf{v} \cdot \nabla)\tilde{n} + (\mathbf{g} \cdot \nabla_v)\tilde{n} + (\tilde{\mathbf{g}} \cdot \nabla_v)\Psi = -\nabla_v \cdot \{\tilde{n}\tilde{\mathbf{g}} - \langle \tilde{n}\tilde{\mathbf{g}} \rangle\},$$

$$(6.14) \quad \nabla \cdot \tilde{\mathbf{g}} = -4\pi G m_* \int \tilde{n} \, d\tau(\mathbf{v}).$$

From (6.13) and (6.14), one may visualize the construction of equations governing the correlation $\langle \tilde{n}\tilde{\mathbf{g}} \rangle$, which appeared in (6.11). Clearly, this would introduce higher correlations, and one arrives at an infinite hierarchy of equations, reminiscent of the well-known procedure used in the theory of turbulence and on other occasions.

An approximation procedure must be decided upon. The simplest is to neglect the right-hand side of (6.13), on the assumption that this would not seriously compromise the evaluation of $\langle \tilde{n}\tilde{\mathbf{g}} \rangle$, which appears on the right-hand side of (6.11).

It is clear that, even after this approximation, one has to integrate a linear partial differential equation in time and in phase space variables. Wu has been able to carry the analysis to a point where formulas can be given for the coefficients of "dynamic friction" and "diffusion in velocity space," which appear in Chandrasekhar's generalized Fokkar-Planck equation, in terms of the elementary modes (*collective modes*) of the linearized form of the partial differential equation (6.13). These modes can be worked out in a certain special case of homogeneous distribution.

The analysis carried out so far clarifies the role of the collective modes, the modes of the linear equation mentioned before. It also shows that Chandrasekhar's original theory yields excellent results. In terms of the present formalism, it amounts to an approximate evaluation of the right-hand side of (6.11) by using the concept of binary collisions.

7. GLOBULAR CLUSTERS AND DENSE GALACTIC NUCLEI

Let us now briefly consider the dynamics of some specific systems in which stellar collisions are important: globular clusters and dense galactic nuclei.

Globular clusters. Since there is a sufficiently rapid relaxation process in a globular cluster, it will presumably tend to evolve into some state of equilibrium. Offhand, one would think of this as a state of *maximum entropy* and hence an *"isothermal"* state. The stellar velocities would follow the Maxwell-Boltzmann distribution, and the principle of equipartition of energy holds. The mass distribution in such a cluster would be identical with that of an isothermal self-gravitating gaseous sphere.

The solution for the density distribution of such a sphere depends on the integration of a partial differential equation of the form

$$(7.1) \qquad \nabla^2\chi = e^{-\chi},$$

where χ is proportional to the gravitation potential. Such calculations have been made numerically and the results found satisfactory in comparison with observations (see Chandrasekhar [2], King [17]). It has also been found that such an isothermal sphere contains an infinite amount of matter, and can therefore not be a completely realistic model for a globular cluster. As one might expect, the stars in the outer parts of the isothermal model are only loosely attached to the globular cluster and must therefore be easily torn away by its interaction with other objects or the action of the general galactic field. There must be *escape* of stars from the globular cluster. This part of the cluster is also so rarefied, that the relaxation process is slow; thus, statistical equilibrium cannot be expected to be attained. However, one might expect the isothermal model to be essentially correct for the interior parts of globular clusters, where most of the mass is concentrated. Indeed, even here the Maxwellian distribution of velocities would imply that the "tail" part represents stars which can escape to infinity.

The rate of escape of stars has been studied in detail by many workers (see the recent survey by Michie [37]). The real phenomenon is remarkably complex, since it involves "diffusion" in the velocity space as well as the configuration space. In one simple model advanced by Spitzer and Härm [42], in which the actual gravitational potential is approximated by a square well, the rate of escape is given by

$$(7.2) \qquad \frac{1}{N}\frac{dN}{dt} = -\frac{1}{88T_R},$$

where N is the total number of stars in the cluster, and T_R is a relaxation time scale within the cluster, given by[†]

$$(7.3) \qquad T_R = \left(\frac{2}{3}\right)^{1/2} \frac{v_s^3}{3\pi G^2 m_*^2 n_0 \log (N/2)},$$

where n_0 is the number of stars per unit volume, m_* is the stellar mass and v_s is the root-mean-square relative velocity in three dimensions.

Galactic nuclei. As mentioned before, a normal spiral galaxy usually has a *nearly* spherical galactic nucleus. The evolution of such a nucleus, especially when it is very dense, has been studied recently by Lyman Spitzer and his collaborators (Spitzer and Saslaw [43], Spitzer and Stone [44]). The dynamics of the stellar system is essentially the same as that for a globular cluster, except that there is a fairly strong coupling with the galactic disk and there is also a fair amount of rotation. But a principal stimulus for the study of these nuclei is the possibility that "at least some of the very luminous radio sources may be galactic nuclei," which are going through a certain stage of the evolution process at the present time. "The energies radiated at the luminosity peak are similar to those observed from quasi-stellar sources, but it is uncertain whether existing nuclei have evolved as yet to the degree of compactness required to account for these highly luminous objects." (Quotations are from Spitzer and Saslaw [43].)

In view of the difficulties in developing the theory precisely when the densities are high, collisions are frequent, and multiple collisions may be important, various investigators have resorted to the use of computing machine methods. I shall not go into detailed discussions of these extensive investigations, hoping that some of these workers will come out with a comprehensive survey.

8. SOME FURTHER REMARKS ON COLLECTIVE MODES

As mentioned before, the spiral waves in galactic disks are collective modes. The nature of these modes has been studied.

[†] Cf. Chandrasekhar [2, Chaps. II and V].

extensively in connection with recent work on electromagnetic plasmas. In particular, it is known that there is the phenomenon of *Landau damping*; that is, these collective modes may dissolve even though there is no close encounter between the particles. Analogous studies in stellar systems have been developed by Lynden-Bell [32], [33] for the case of the homogeneous system in uniform rotation.

Recently, a number of authors, including Hénon [11], King [16], Lecar [21], and Lynden-Bell [34], [35], have discussed the possible role of these collective modes as an agent of "coarse grain" relaxation in phase space, during the formative period of a galaxy. The study was prompted by the following observational evidence which suggests some form of equipartition even in collisionless systems: (a) the light distribution observed in spherical galaxies is found to be in reasonable agreement with that predicted on the basis of modified isothermal spheres, developed by Michie [36] and by King [16]; (b) yet there is no segregation by mass of the stars, as one would predict from a genuine theory of equipartition.

Numerical experiments were recently made by Hénon [11] and by Lecar [21] to follow up this type of process. Lynden-Bell developed a statistical mechanics based on this mechanism. It turns out to be the same as Fermi-Dirac statistics, apart from a constant permutation factor, since the particles are identifiable; yet the exclusion principle still holds.

9. CONCLUDING REMARKS

I have presented a brief survey of the statistical theory of a stochastic process in which the stars are the atoms. As mentioned above, since we ourselves are of subatomic dimensions, we are interested in examining the problem both collectively and individually. The deductions of the mathematical theory can yield results for our own galaxy, which are highly satisfactory from the observational point of view. But there are still many basic problems to be solved, and many observational data to be explained.

At this point, I wish to make two brief remarks of a general nature.

First, in statistical mechanics, combinatorial analysis is extremely important: the problem is discrete. At the same time, when one goes into the consideration of the limiting form of the distribution function, one finds it is continuous (in phase space). Thus, it would be unwise to divide applied mathematics artificially in terms of discrete (or finite) and continuous (or infinite) parts. One often has to consider both and to carry out the limiting process in a single theory that deals with one particular subject.

Second, the current revival in the theoretical study of galactic structure is largely stimulated by the recent improvement in observational technique (especially radio astronomy). In turn, these mathematical studies enable us to understand better the nature of collective modes and to raise again, in sharper focus, some fundamental questions on the relaxation process in a "collisionless" system. These studies certainly involve quite sophisticated mathematical concepts and theorems. They point toward the interest and the importance of the study of *nonlinear random processes*, and might, by example, contribute to the stimulation and the development of a general mathematical theory, even as the study of the physical process of Brownian motion did. It is in the hope that more general mathematical theories will be stimulated by these studies that I wish to dedicate this survey paper to the late John von Neumann. In particular, I hope that the problem of greatest interest to him, the relaxation process in stellar systems, will be put on even firmer foundation in all its generality. One would then also be in a better position to judge whether the mysterious quasi-stellar sources are indeed galactic nuclei, so dense that the stars are literally annihilating each other with explosive release of energy many million or even billion times larger than the most violent explosives ever produced by mankind, now or in the future.

APPENDIX A. COLLECTIVE MODES AND THEIR EFFECTS ON THE RELAXATION PROCESS

Consider a uniformly rotating system of stars at such an angular velocity that the centrifugal forces and the gravitational forces are in balance. Consider a small deviation of the distribution function Ψ from its homogeneous equilibrium condition Ψ_0:

(A1) $$\Psi' = \Psi - \Psi_0.$$

Then the linearized equation governing Ψ' is

(A2) $$\frac{\partial \Psi'}{\partial t} + (\mathbf{v} \cdot \nabla)\Psi' - 2(\Omega \times \mathbf{v}) \cdot \nabla_v \Psi' = \nabla V' \cdot \frac{\partial \Psi_0}{\partial \mathbf{v}},$$

(A3) $$\nabla^2 \Psi' = 4\pi G m_* \int \Psi' \, d\tau(\mathbf{v}).$$

If one looks for solutions of the type

(A4) $$\Psi' = f(\mathbf{k}, \mathbf{v}, \omega)e^{i(\mathbf{k} \cdot \mathbf{r} - \omega t)},$$

(A5) $$V' = \phi(\mathbf{k}, \omega)e^{i(\mathbf{k} \cdot \mathbf{r} - \omega t)},$$

one finds that indeed this is possible provided that the following dispersion relation is satisfied:

(A6) $$0 = 1 + \frac{\omega_0^2}{k^2} \sum_{n=-\infty}^{\infty} \int d\tau(v) F_n(\mathbf{k}, \mathbf{v}; \omega),$$

where

$$F_n(\mathbf{k}, \mathbf{v}; \omega) = \frac{J_n^2(k_\perp v_\perp/(2\Omega))}{k_z v_z + 2n\Omega - \omega}\left(k_z \frac{\partial \Psi_0}{\partial v_z} + \frac{2n\Omega}{v_\perp} \frac{\partial \Psi_0}{\partial v_\perp}\right),$$

and

(A7) $$\omega_0^2 = 4\pi G m_* n = 4\pi G \rho.$$

The symbols v_z and k_z denote the components of velocity and of wave vector in the direction parallel to the axis of rotation, which is taken to be the z-axis. The symbols $v_\perp$ and $k_\perp$ denote the (full) components of velocity at right angles to the z-axis.

The integral in v_z is defined for positive values of the imaginary part of ω. Analytic continuation must be resorted to when $\mathrm{Im}(\omega) < 0$, for there is a pole at $v_z = (\omega - 2n\Omega)/k_z$.

Although the general form of (A6) appears quite unmanageable, it has been found possible to discuss the special cases (a) $k_\| = 0$, i.e., waves propagating along the z-axis, and (b) $k_\perp = 0$, i.e., waves whose normal is at right angles to the z-axis. In the former case, one finds instability for wavelengths λ larger than a critical length λ_J, obtained earlier by Jeans on the basis of a gas dynamic theory, and damping of the waves (according to Landau's concept) when $\lambda < \lambda_J$. In the gas dynamical case, one finds neutral waves.

APPENDIX B. SOLUTION OF A PARTIAL DIFFERENTIAL EQUATION OF A CERTAIN TYPE

Consider the differential equation

(B1) $a(x, y)\psi_x + b(x, y)\psi_y = if(x, y)\psi + g(x, y) + ih(x, y),$

where $a(x, y), b(x, y), \cdots, h(x, y)$ are all real. Consider the cases where the characteristic curves

(B2) $$\frac{dx}{a} = \frac{dy}{b}$$

are *closed curves* around the origin, which is a singularity of the differential equation (B2). The singularity is then said to be a *center* (see Coddington and Levinson [6, pp. 371, 375]). In our present problem, the existence of such integral curves is guaranteed

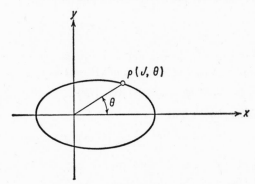

Fig. 5.

by the existence of the integrals of angular momentum and energy. Thus, we have, along a characteristic curve (see Fig. 5),

(B3) $x = \rho(J_1, J_2, \theta) \cos \theta,$

$y = \rho(J_1, J_2, \theta) \sin \theta.$

The characteristic curve of (B1) in the three-dimensional (a, b, ψ) space is given by

$$\frac{dx}{a} = \frac{dy}{b} = \frac{d\psi}{if\psi + g + ih}.$$

By introducing (B3) we may write the differential equation for ψ as

(B4) $$\frac{d\psi}{d\theta} = if_1(\theta, J)\psi + g_1(\theta, J),$$

where

(B5) $$f_1 = \frac{f}{a}\frac{dx}{d\theta}, \qquad g_1 = \frac{g + ih}{b}\frac{dy}{d\theta},$$

and both are expressed as function of θ by means of (B3), with J_1 and J_2 (denoted collectively by J) as parameters. The integration of the differential equation (B4) will not be considered.

We observe that the functions $f_1(\theta, J)$ and $g_1(\theta, J)$ are derived from single-valued functions of (x, y) and are therefore periodic functions of the angular variable θ (cf. Fig. 5). We also want only single-valued solutions of (B1), and hence $\psi(\theta, J)$ must also be periodic in θ (with period 2π).

THEOREM: *Consider the differential equation*

(B6) $$\frac{d\psi}{d\theta} = if_1(\theta, J)\psi + g_1(\theta, J),$$

where f_1 and g_1 are periodic functions of θ (with period 2π), f_1 is real, and J is a parameter. The solution of (B6) is uniquely given by

(B7) $$\psi = (\Phi e^{iv\theta}) \exp \{iF\},$$

where $v(J)$ is so chosen that

(B8) $$F(\theta, J) = \int_0^\theta \{f_1(\theta, J) - v(J)\} \, d\theta$$

is a periodic function of θ (with period 2π), and $\Phi e^{iv\theta}$ is another periodic function of θ defined as follows:

(B9) $\Phi e^{iv\theta}$

$$= (1 - e^{2\pi iv})^{-1} \int_0^{2\pi} g_1(\theta - \varphi, J) \exp \{iv\varphi - F(\theta - \varphi, J)\} \, d\varphi.$$

[Note that $v = v(J)$ when we consider (B7) as a solution of (B1).]
To solve the equation

(B10) $$\frac{d\psi}{d\theta} = if_1(\theta, J)\psi + g_1(\theta, J),$$

consider first the solution of the homogeneous equation

(B11) $$\frac{d\psi_0}{d\theta} = if_1(\theta, J)\psi_0,$$

and then attempt to solve (B10) by writing

(B12) $$\psi = \psi_0\Phi.$$

We want to obtain the solution so that all the functions appear in periodic form.

Now, (B11) can be easily solved for ψ_0 in the form

(B13) $$\psi_0 = \exp\left\{i\int f_1(\theta, J)\, d\theta\right\}.$$

This is in general not a periodic function of θ. However, we may exhibit the periodic part by introducing the parameter $v(J)$ defined by

(B14) $$\int_0^{2\pi} f_1(\theta, J)\, d\theta = 2\pi v(J).$$

The function

(B15) $$F(\theta, J) = \int_0^\theta \{f_1(\theta', J) - v\}\, d\theta'$$

is then a periodic function of θ with period 2π. Thus, the solution (B13) can be written in the more convenient form

(B16) $$\psi_0 = e^{iv\theta} \exp\{iF(\theta, J)\},$$

where the second factor is periodic.

To obtain Φ in (B12), we substitute ψ from (B12) into (B10) and obtain

(B17) $$\Phi = \int_0^\theta g_1(\theta', J)\{\psi_0(\theta', J)\}^{-1}\, d\theta' + C(J).$$

We shall determine $C(J)$ in such a manner that (B12) is periodic in θ with period 2π; i.e., $\Phi(\theta, J)e^{iv\theta}$ is periodic in θ with period 2π. This is done as follows.

When $\Phi e^{iv\theta}$ is written out explicitly by the use of (B16) and (B17), the transformation

(B18) $$\theta - \theta' = \varphi$$

suggests itself. We then find that

(B19) $$\Phi e^{iv\theta} = \Phi_1(\theta) + C(J)e^{iv\theta},$$

where

(B20) $\Phi_1(\theta) = \int_0 g_1(\theta - \varphi, J) \exp\{iv\varphi - iF(\theta - \varphi, J)\}\, d\varphi.$

Let us examine the behavior of all the functions as θ changes to $\theta + 2\pi$. Clearly, if $\Phi e^{iv\theta}$ is to take on the same value, we must have

(B21) $\Phi_1(\theta + 2\pi) - \Phi_1(\theta) = C(J)e^{iv\theta}(1 - e^{2\pi iv}).$

If we now introduce the new variable $\varphi' = \varphi - \theta$, the function on the left-hand side of (B21) is, by (B20),

$$\int_\theta^{\theta+2\pi} g_1(\theta = \varphi, J) \exp\{iv\varphi - iF(\theta - \varphi, J)\}\, d\varphi$$

$$= \int_0^{2\pi} g_1(-\varphi', J) \exp\{iv(\varphi' + \theta) - iF(-\varphi', J)\}\, d\varphi'.$$

Thus,

(B22) $\Phi_1(\theta + 2\pi) - \Phi_1(\theta)$

$$= e^{iv\theta} \int_0^{2\pi} g_1(-\varphi', J) \exp\{iv\varphi' - iF(-\varphi', J)\}\, d\varphi',$$

and we must choose

(B23) $C(J)$
$$= (1 - e^{2\pi iv})^{-1} \int_0^{2\pi} g_1(-\varphi', J) \exp\{iv\varphi' - iF(-\varphi', J)\}\, d\varphi'.$$

We may also transform $F(\theta + 2\pi)$ in another manner to obtain the explicit form (B9) for Φ. For this purpose, let us write

(B24) $\Phi_1(\theta + 2\pi) = \{\Phi_1(2\pi) - \Phi_1(0)\} + \{\Phi_1(2\pi + \theta) - \Phi_1(2\pi)\},$

where the differences are to be thought of as integrals with integrand given by (B20). In the second integral, introduce the new variable $\varphi'' = \varphi - 2\pi$, and we obtain

(B25) $\Phi_1(\theta + 2\pi) = \{\Phi_1(2\pi) - \Phi_1(0)\} + e^{2\pi iv}\Phi_1(\theta).$

Combining (B21) and (B25), we obtain

(B26) $\{\Phi_1(\theta) + C(J)e^{iv\theta}\}(1 - e^{2\pi iv}) = \Phi_1(2\pi) - \Phi_1(0),$

which gives immediately the form (B9), by (B19) and (B20).

APPENDIX C. SHU'S SOLUTION OF STELLAR RESPONSE TO SPIRAL WAVES

Shu calculated the response of a stellar to a spiral gravitational field by the introduction of the energy integral E and the angular momentum integral J as independent variables in place of the components of velocity dispersion in the plane of the disk. In this way, the mathematical manipulation is simplified, and the analysis can be carried through for conditions near Lindblad resonance. His result is herewith quoted.

Let the basic distribution be represented by

$$(C1) \qquad \Psi_0 = F_0(E_0, J),$$

where E_0 is the energy integral for the undisturbed symmetrical potential $V_0(\bar{\omega})$. Let the resultant potential be

$$(C2) \qquad \begin{aligned} V(\bar{\omega}, \theta, t) &= V_0(\bar{\omega}) + V_1(\bar{\omega}, \theta, t) \\ &= V_0(\bar{\omega}) + \hat{V}(\bar{\omega}) \exp\{i(\omega t - m\theta)\} \end{aligned}$$

in the modified form; and let us denote the modified distribution function by

$$(C3) \qquad \Psi = F_0(E, J) + F_1(\bar{\omega}, \theta, E, J, t).$$

To a first approximation we then have

$$(C4) \qquad \Psi = \Psi_0 + \frac{\partial F_0}{\partial E_0} V_1(\bar{\omega}, \theta, t) + F_1(\bar{\omega}, \theta, E_0, J, t).$$

(Notice the change from E to E_0.) Shu found that if we now write

$$(C5) \qquad F_1(\bar{\omega}, \theta, E_0, J, t) = f(\bar{\omega}, E_0, J) \exp\{i(\omega t - m\theta)\},$$

then

$$(C6) \quad f(\bar{\omega}, E_0, J)$$
$$= -\frac{\omega \partial F_0/\partial E_0 + m \partial F_0/\partial J}{2 \sin(\omega \tau_0 - m\theta_0)} \int_{-\tau_0}^{\tau_0} \hat{V}(\bar{\omega}_0(\tau)) e^{i(\omega\tau - \bar{\theta}_0(\tau))} d\tau,$$

where $\{\bar{\omega}_0(\tau), \bar{\theta}_0(\tau)\}$ gives the trajectory of a star in the undisturbed disk, with specified values of E_0 and J, such that

$$(C7) \qquad \bar{\omega}_0(\pm\tau_0) = \bar{\omega}$$

and

(C8) $\theta_0(\pm\tau_0) = \pm\theta_0.$

These results, when expanded into the asymptotic form for tightly wound spirals, agree with those obtained by direct calculation. For conditions near resonance, they furnish valid results while the asymptotic method fails.

REFERENCES

1. Allen, C. W., *Astrophysical Quantities*. London: The Athlone Press, 1963.

2. Chandrasekhar, S., *Principles of Stellar Dynamics*. Chicago: The University of Chicago Press, 1942, and New York: Dover Publications, 1960.

3. ———, "The equilibrium and the stability of Riemann ellipsoids, I," *Astrophys. J.*, **142** (1965), 890–921.

4. ———, "The equilibrium and the stability of the Riemann ellipsoids, II," *Astrophys. J.*, **145** (1966), 842–877.

5. Chandrasekhar, S., and John von Neumann, "The statistics of the gravitational field arising from a random distribution of stars," *Astrophys. J.*, **95** (1942), 489–531.

6. Coddington, E. A., and N. Levinson, *Theory of Ordinary Differential Equations*. New York: McGraw-Hill, 1955.

7. Contopoulos, G., "Applications of the third integral in the galaxy," *Relativity Theory and Astrophysics, Part 2, Galactic Structure*, J. Ehlers, ed. Providence, R.I.: American Mathematical Society, 1967.

8. Freeman, K. C., "Gas streaming in barred galaxies," *Monthly Notices Roy. Astronom. Soc.*, **130** (1965), 183–197.

9. ———, "Structure and evolution of barred spiral galaxies, *I, II, III*," *Monthly Notices Roy. Astronom. Soc.*, **133** (1964), 47–62; **134** (1966), 1–14, 15–23.

10. Freeman, K. C., and L. Mestel, "On magnetic field and gas streaming in spiral galaxies," *Monthly Notices Roy. Astronom. Soc.*, **134** (1966), 37–51.

11. Hénon, M., "L'évolution initiale d'un amas sphérique," *Ann. Astrophys.*, **27** (1964), 83–91.

12. James, R. A., "The structure and stability of rotating gas masses," *Astrophys. J.*, **140** (1964), 552–582.

13. Jeans, J. H., *Astronomy and Cosmogony.* Cambridge, England: University Press, 1929.

14. King, I. R., "The structure of star clusters. I. An empirical density law," *Astronom. J.*, **67** (1962), 471–485.

15. ———, "The structure of star clusters. II. Steady state velocity distributions," *Astronom. J.*, **70** (1965), 376–383.

16. ———, "The structure of star clusters. III. Some simple dynamical models," *Astronom. J.*, **71** (1966), 64–75.

17. ———, "The dynamics of star clusters," *Relativity Theory and Astrophysics, Part 2, Galactic structure*, J. Ehlers, ed. Providence, R.I.: American Mathematical Society, 1967, 116–130.

18. Klimontovich, Yu. L., *Statistical Theory of Nonequilibrium Processes in a Plasma.* Moscow: Moscow University Press, 1964.

19. Lebovitz, N. R., "On Riemann's criterion for the stability of liquid ellipsoids," *Astrophys. J.*, **145** (1966), 878–885.

20. ———, "Rotating fluid masses," *Annual Rev. Astronom. Astrophys.*, **5** (1967), 465–480.

21. Lecar, M., "One-dimensional self-gravitating stellar gas," *International Astronomical Union Symposium No. 25.* New York: Academic Press, 1966, 46–48.

22. Lin, C. C., "On the mathematical theory of a galaxy of stars," *J. SIAM Appl. Math.*, **14** (1966), 876–921.

23. ———, "Stellar dynamical theory of normal spirals," *Relativity Theory and Astrophysics, Part 2, Galactic Structure*, J. Ehlers, ed. Providence, R.I.: American Mathematical Society, 1967, 66–97.

24. ———, "The dynamics of disk-shaped galaxies," *Annual Rev. Astronom. Astrophys.*, **5** (1967), 453–464.

25. ———, "Spiral structure in galaxies," *Galaxies and the Universe*, L. Woltjer, ed. New York: Columbia University Press, 1968, pp. 33–51.

26. Lin, C. C., and F. H. Shu, "On the spiral structure of disk galaxies," *Astrophys. J.*, **140** (1964), 646–655.

27. Lin, C. C., and F. H. Shu, "On the spiral structure of disk galaxies. II. Outline of a theory of density waves," *Proc. Nat. Acad. Sci.*, **55** (1966), 229–234.

28. ———, "Density waves in disk galaxies," *Proc. International Astronomical Union-URSI Symp. No. 31.* Noordwijk, 1966, pp. 313–317.

29. Lin, C. C., C. Yuan, and F. H. Shu, "On the spiral structure of disk galaxies. III. Comparison with observations," *Astrophys. J.*, **155**, 721.

30. Lindblad, B., "On the possibility of a quasi-stationary spiral structure in galaxies," *Stockholm Obs. Ann.*, **22** (1963), 3–20.

31. Lynden-Bell, D., "Can spherical clusters rotate?" *Monthly Notices Roy. Astronom. Soc.*, **120** (1960), 204–213.

32. ———, "Stellar dynamics: Potentials with isolating integrals," *Monthly Notices Roy. Astronom. Soc.*, **124** (1962), 95–123.

33. ———, "Stellar dynamics: The stability and vibrations of a gas of stars," *Monthly Notices Roy. Astronom. Soc.*, **124** (1962), 279–296.

34. ———, "Statistical mechanics of violent relaxation in stellar systems," *Monthly Notices Roy. Astronom. Soc.*, **136** (1967), 101–121.

35. ———, "Cooperative phenomena in stellar dynamics," *Relativity Theory and Astrophysics, Part 2, Galactic Structure*, J. Ehlers, ed. Providence, R.I.: American Mathematical Society, 1967, pp. 131–168.

36. Michie, R. W., "On the distribution of high energy stars in spherical stellar systems," *Monthly Notices Roy. Astronom. Soc.*, **125** (1963), 127–139.

37. ———, "The dynamics of star clusters," *Annual Rev. Astronom. Astrophys.*, **2** (1964), 49–72.

38. Milder, D. M., "Dynamics of flattening in rotating stellar systems." Doctoral thesis, Department of Physics, Harvard University, Cambridge, 1967.

39. Ng, E. W., "Self-consistent models of disk galaxies," *Astrophys. J.*, **150** (1967), 787–796.

40. Prendergast, K. H., "The motion of gas in barred spiral galaxies," *Interstellar Matter in Galaxies*, L. Woltjer, ed. New York: W. A. Benjamin, pp. 217–221.

41. Shu, Frank H., "The dynamics and large-scale structure of spiral galaxies." Doctoral thesis, Department of Astronomy, Harvard University, Cambridge, 1968.

42. Spitzer, L., and R. Härm, "Evaporation of stars from isolated clusters," *Astrophys. J.*, **127** (1958), 544–550.

43. Spitzer, L., and W. C. Saslaw, "On the evolution of galactic nuclei," *Astrophys. J.*, **143** (1966), 400–419.

44. Spitzer, L., and M. E. Stone, "On the evolution of galactic nuclei, II." *Astrophys. J.*, **147** (1967), 519–528.

45. Toombe, A., "On the gravitational stability of a disk of stars," *Astrophys. J.*, **139** (1964), 1217–1235.

46. Vandervoort, P. O., "The equilibrium of rapidly rotating galaxies," *Astrophys. J.*, **147** (1967), 91–111.

47. Wu, C. S., "Kinetic theory of a self-gravitating system with uniform rotation," *Phys. Fluids*, **11** (1968), Astrophys. J., 316–325.

RELATIVISTIC HYDRODYNAMICS

A. H. Taub

1. INTRODUCTION

Prerelativity hydrodynamics, like all of Newtonian continuum mechanics, is concerned with motion of media in a three-dimensional Euclidean space and uses the notions of absolute simultaneity inherent in Newtonian mechanics. Thus prerelativity hydrodynamics may be formulated in terms of four independent variables which may be taken as the three cartesian coordinates of a point in Euclidean three-space x^i ($i = 1, 2, 3$) and the time t. The subject is concerned with the description of the state of a fluid or gas as a function of these four independent variables. The state is described macroscopically by five dependent variables, which may be taken to be two thermodynamic ones, such as the pressure p and the density ρ, and three kinematic ones such as the components of the Eulerian velocity field (the velocity of that part of the medium which is at a certain place at a certain time). The kind of fluid or gas with which one is dealing may be described by specifying its caloric equation of state, that is, by prescribing the specific

150

internal energy ϵ as a function of pressure and density. Thus for
a compressible gas one has

$$\epsilon = \frac{1}{\gamma - 1}\, p/\rho$$

when γ is the ratio of specific heats of the gas.

Since changes in the state of a given element of the medium
must be in accordance with the laws of conservation of mass,
momentum, and energy, and since these laws relate the changes of
state of one element of the medium with those of its neighbors, the
subject of hydrodynamics deals with solutions of the five equations
describing these conservation laws which satisfy the initial conditions and the boundary conditions.

For an ideal fluid discontinuities in the state variables may occur
even for continuous initial and boundary values. This means that
one cannot use the partial differential equations embodying the
conservation laws throughout the domain occupied by the fluid
but must supplement these with various algebraic statements
(called the Rankine-Hugoniot conditions and derived from the
conservation laws) relating the jumps in the state variables across
the discontinuities. Thus a general hydrodynamic problem involves the patching together of various solutions of the partial
differential equations, each holding in different domains, across
unknown moving surfaces of discontinuity in such a way that the
Rankine-Hugoniot equations are satisfied.

As is well-known, one type of discontinuity that may occur is a
shock whose normal velocity depends on the jump in pressure.
This velocity increases as the jump in pressure increases and may
take on an arbitrarily large value for a suitable choice of the jump
in pressure.

The equations embodying the conservation laws are invariant
under the galilean group of transformations. If the Eulerian
coordinates described above are used, the equations may be shown
to be numerically invariant under the transformation

$$x^{*i} = x^i - u^i t \qquad (i = 1, 2, 3),$$

where u^i are three constants, as well as under transformations of cartesian coordinates. That is, if we replace all velocity vectors with components v^i by vector with components $v^i - u^i$, x^i by x^{*i} and leave p and ρ unaltered, the equations embodying the conservation laws remain unaltered.

The five conservation laws of prerelativity hydrodynamics are derivable from kinetic theory. In the latter theory the Maxwell-Boltzmann distribution function plays a crucial role. It is usually considered as a function of seven variables: the time, the coordinates x^i of a point in space, and the components v^i of a velocity vector relative to the coordinate system used in the description of Euclidean three-space. The domain of the latter three variables in cartesian coordinates is

$$-\infty \leq v^i \leq \infty.$$

It is evident that if kinetic theory is to be modified to be consistent with the postulates of the special theory of relativity, a change must be made, since these postulates do not allow particle velocities to exceed the velocity of light. The change required is to replace the variables v^i by the momentum per unit rest mass. In classical theory these two quantities are proportional; in special relativity in a galilean frame we may write

$$(1.1) \quad \xi_i = \frac{v_i}{(1 - v^2/c^2)^{1/2}}, \qquad v_i = \frac{\xi_i}{(1 - \xi^2/c^2)^{1/2}}, \qquad i = 1, 2, 3,$$

where

$$v^2 = \sum_{i=1}^{3} v_i^2, \qquad \xi^2 = \sum_{i=1}^{3} \xi_i^2.$$

Then, if the v_i are the components of the velocity of a particle relative to an inertial coordinate system in Minkowski space-time, the ξ_i are the components of the momentum per unit rest mass of this particle relative to the same coordinate system. Let $f(t, x^i, \xi_i)$ be the number of particles in the region x^i to $x^i + dx^i$ with values of ξ_i between ξ_i and $\xi_i + d\xi_i$ at time t in the chosen coordinate system. The Boltzmann equation for f is then

$$(1.2) \quad Df = \frac{\partial f}{\partial t} + \frac{\xi_i}{(1 + \xi^2/c^2)^{1/2}} \frac{\partial f}{\partial x^i} + F_i \frac{\partial f}{\partial \xi_i} = \Delta_c(f)$$

where F_i is the external force per unit mass and $\Delta_c(f)$ is the time rate of change of f due to collisions between the particles.

Let x stand for the four coordinates x^1, x^2, x^3, and t.

If $f(x, \xi_i)$, $f'(x', \xi_i')$ are the distribution functions of a gas with respect to two different inertial coordinate systems and if x, x' and ξ_i, ξ_i' belong to the same event and the same four-momentum, respectively, then $f(x, \xi_i) = f'(x', \xi_i')$; in this sense the distribution function is a scalar. In fact, the equation

$$\frac{d_3\xi}{(1 + \xi^2/c^2)^{1/2}} = \frac{d_3\xi'}{(1 + \xi'^2/c^2)^{1/2}}$$

expresses that the parameter-intervals $d_3\xi$, $d_3\xi'$ describe the same infinitesimal domain of four-momenta, and

$$(1 + \xi^2/c^2)^{1/2} d_3x = (1 + \xi'^2/c^2)^{1/2} d_3x'$$

holds if d_3x, d_3x' are the volumes of two cross sections $t = $ const., $t' = $ const., respectively, of a world tube swept out in space-time by a cloud of particles with four-momenta close to $m(\xi_i, (1 + \xi^2/c^2)^{1/2})$ (Lorentz-contraction). Hence, $d_3x\, d_3\xi = d_3x'\, d_3\xi'$ holds under conditions where the same particles "belong to" $d_3x\, d_3\xi$ and to $d_3x'\, d_3\xi'$, which proves our assertion; cf. [13].

In terms of the function f we may define the following functions of x_i and t:

$$(1.3) \quad U^\mu(x) = \int V^\mu(\xi) f(x, \xi_i) \frac{d_3\xi}{(1 + \xi^2/c^2)^{1/2}} = \int V^\mu(\xi)\, d\mu(\xi)$$

and

$$(1.4) \quad T^{\mu\nu}(x) = mc^2 \int V^\mu(\xi) V^\nu(\xi) f(x, \xi) \frac{d_3\xi}{(1 + \xi^2/c^2)^{1/2}} = T^{\mu\nu}(x)$$

$$= mc^2 \int V^\mu(\xi) V^\nu(\xi)\, d\mu(\xi), \qquad \mu, \nu = 1, 2, 3, 4,$$

where

$$V^i = \xi_i/c, \qquad V^4 = (1 + \xi^2/c^2)^{1/2},$$

and hence

$$(1.5) \quad g_{\mu\nu} V^\mu V^\nu = (V^4)^2 - \sum_{i=1}^{3} (V^i)^2 = 1,$$

and the integrations are carried out over the entire volume of ξ_1, ξ_2, ξ_3-space ($-\infty \leq \xi_i \leq \infty$).

Since, as remarked above, $d_3\xi/(1 + \xi^2/c^2)^{1/2}$ is Lorentz-invariant, and f is a scalar and $V^\mu(\xi)$ are four-vectors, it follows from these definitions that $U^\mu(x)$ is a four-vector field and that $T^{\mu\nu}(x)$ is a tensor field in space-time.

We next assume that the laws of collision are such that

(1.6) $$\int \phi^\alpha \Delta_c(f) \, d_3\xi = 0, \qquad \alpha = 0, 1, 2, 3, 4,$$

where

$$\phi^0 = m, \qquad \phi^i = m\xi_i, \qquad \phi^4 = m(1 + \xi^2/c^2)^{1/2}.$$

This assumption is the special relativistic equivalent of the assumption that the total rest mass, momentum, and energy of the collection of the particles in unit volume at x is conserved in the collisions.

In case $\alpha = 0$ in Eq. (1.6) it follows that we must have

$$m \int Df d_3\xi = m \int \left(\frac{\partial f}{\partial t} + \frac{\xi_i}{(1 + \xi^2/c^2)^{1/2}} \frac{\partial f}{\partial x^i} + F_i \frac{\partial f}{\partial \xi_i} \right) d_3\xi = 0,$$

or

(1.7) $$m(\partial U^4/\partial t + cU^i \partial f/\partial x^i) = mc(U^\alpha)_{,\alpha} = 0,$$

where

$$f_{,i} = \frac{\partial f}{\partial x^i}, \qquad f_{,4} = \frac{1}{c} \frac{\partial f}{\partial t}.$$

We shall define the scalar rest mass density of the fluid by the equation

$$\rho^2 = m^2 U^\alpha U_\alpha$$

and write

$$U^\alpha = \frac{\rho}{m} u^\alpha.$$

Then

$$u^\alpha u_\alpha = 1,$$

and Eq. (1.7) becomes

(1.8) $$(\rho u^\mu)_{,\mu} = 0.$$

This equation is known as the equation of conservation of mass in classical hydrodynamics. It is often referred to as the equation of conservation of particle number.

In case $\alpha = i$ in Eq. (1.6) it follows that we must have

(1.9) $\displaystyle \int m\xi_i Df \, d_3\xi$

$$= m \int \left(\xi_i \frac{\partial f}{\partial t} + \frac{\xi_i \xi_j}{(1 + \xi^2/c^2)^{1/2}} \frac{\partial f}{\partial x^j} + \xi_i F_j \frac{\partial f}{\partial \xi^j} \right) d_3\xi$$

$$= T^{i\mu}_{,\mu} - mU^4(x)F_i = 0.$$

The last term in the last equation arises from the corresponding term in the next to last equation by an integration by parts. Equation (1.9) and the equation arising from Eq. (1.6) by setting $\alpha = 4$ may be written as

(1.10) $$T^{\mu\nu}_{,\nu} = \rho\mathfrak{F}^\mu,$$

where

$$\mathfrak{F}^i = \frac{F_i}{(1 - u^2/c^2)^{1/2}}, \qquad \mathfrak{F}^4 = \frac{F_i u^i}{c}$$

is the four-dimensional force vector per unit rest mass.

We shall consider Eq. (1.8) and (1.10) as the laws of hydrodynamics in special relativity. In an arbitrary coordinate system in Minkowski space-time and when there are no external forces present they may be written as

(1.11) $$(\rho u^\mu)_{;\mu} = 0$$

and

(1.12) $$T^{\mu\nu}_{;\nu} = 0.$$

There are five equations, corresponding to the five conservation conditions described by Eq. (1.6).

2. DECOMPOSITION OF $T^{\mu\nu}$

The tensor $T^{\mu\nu}$ may be expressed in terms of the timelike unit vector u^μ as

(2.1) $$T^{\mu\nu} = w u^\mu u^\nu + W^\mu u^\nu + W^\nu u^\mu + W^{\mu\nu},$$

where

$$w = u^\mu u^\nu T_{\mu\nu},$$

(2.2) $$W^\mu = T_{\rho\nu} u^\nu (g^{\rho\mu} - u^\rho u^\nu) = T_{\rho\nu} u^\nu h^{\rho\mu}$$

$$W^{\mu\nu} = T_{\rho\sigma}(g^{\rho\mu} - u^{\rho}u^{\mu})(g^{\sigma\nu} - u^{\sigma}u^{\nu}) = T_{\rho\sigma}h^{\rho\mu}h^{\sigma\nu},$$

and hence

(2.3) $$W^{\mu}u_{\mu} = W^{\mu\nu}u_{\nu} = 0.$$

The vector W^{μ} is known as the heat flow-vector, and the tensor $W^{\mu\nu}$ is known as the stress-tensor.

It follows from Eq. (2.1) that

(2.4) $$T = g_{\mu\nu}T^{\mu\nu} = w - 3p,$$

where

(2.5) $$-3p = W^{\mu\nu}g_{\mu\nu}.$$

The scalar function p is called the hydrostatic pressure.

We define the rest specific internal energy of the fluid by the equation

$$w = \rho(c^2 + \epsilon).$$

It is then a consequence of Eqs. (2.2), (1.3), (1.4), and (1.8) and of Schwarz's inequality that ϵ when considered as a function of ρ and p must satisfy the inequality

(2.6) $$\epsilon \geqq \frac{3}{2}\frac{p}{\rho} + c^2\left[\left(1 + \frac{9}{4}\left(\frac{p}{\rho c^2}\right)^2\right)^{1/2} - 1\right].$$

The proof of the inequality (2.6) is as follows: Schwarz's inequality states that

$$\left(\int g(\xi)\,d\mu(\xi)\right)^2 \leqq \left(\int g^2(\xi)\,d\mu(\xi)\right)\left(\int d\mu(\xi)\right)$$

for functions $g(\xi)$ for which the integrals exist. Now define $g(\xi)$ by the equation

$$g(\xi) = V^{\mu}(\xi)U_{\mu}.$$

Hence

$$\int g(\xi)\,d\mu(\xi) = U^{\mu}U_{\mu} = \frac{\rho^2}{m^2}.$$

Further, it follows from Eqs. (1.4) and (1.5) that

$$w - 3p = T = mc^2\int g_{\mu\nu}V^{\mu}V^{\nu}\,d\mu = mc^2\int d\mu$$

and that

$$T_{\mu\nu}U^{\mu}U^{\nu} = mc^2\int g^2(\xi)\,d\mu(\xi) = \frac{\rho^2}{m^2}\,w.$$

Hence we must have

$$\rho^2 c^4 \leqq (w - 3p)w;$$

that is, the inequality (2.6) must hold.

In classical hydrodynamics, the specific internal energy ϵ is written as

$$(2.7) \qquad \epsilon = \frac{1}{\gamma - 1} p/\rho.$$

If we consider Eq. (2.7) as defining γ, no longer a constant, we may write (2.6) as

$$(2.8) \qquad \gamma \leqq 1 + \frac{2x/3}{x + \frac{2}{3}((1 + \frac{9}{4}x^2)^{1/2} - 1)},$$

where

$$(2.9) \qquad x = \frac{p}{\rho c^2} \geqq 0.$$

For x small compared to one the right-hand side of the inequality (2.8) is approximately $\frac{5}{3}$, whereas for x large compared to one it is approximately $\frac{4}{3}$.

The stress tensor $W^{\mu\nu}$ may be further decomposed into

$$W^{\mu\nu} = -ph^{\mu\nu} - P^{\mu\nu},$$

where

$$h_{\mu\nu}P^{\mu\nu} = 0$$

and p is the pressure as defined by equation (2.5). In case the distribution function is the relativistic generalization of the Maxwell function, $P^{\mu\nu} = W^\mu = 0$ and $T^{\mu\nu}$ reduces to that of a perfect fluid.

$$(2.10) \qquad T^{\mu\nu} = wu^\mu u^\nu - ph^{\mu\nu} = (w + p)u^\mu u^\nu - pg^{\mu\nu}.$$

It follows from this equation and the fact that u^μ is a unit vector that

$$T^\mu_\nu u^\nu = wu^\mu,$$

and that

$$T^\mu_\nu X^\nu = -pX^\nu$$

if

$$u_\nu X^\nu = 0.$$

That is, the stress energy tensor of a perfect fluid may be characterized algebraically by the following properties: (1) it has a single timelike vector as a proper vector with a positive proper value $\rho(c^2 + \epsilon)$, and (2) it has three spacelike proper vectors, each with the same proper value $-p$. Such stress energy tensors satisfy the equation

$$(2.11) \qquad (T^\mu_\nu + p\delta^\mu_\nu)(T^\nu_\lambda - \rho(c^2 + \epsilon)\delta^\nu_\lambda) = 0,$$

where

$$
\begin{aligned}
p &= -T/4 + (12S - 3T^2)^{1/2}/12, \\
\rho(c^2 + \epsilon) &= T/4 + (12S - 3T^2)^{1/2}/4 \\
(2.12) \qquad T &= T^\mu_\mu, \\
S &= T^\mu_\rho T^\rho_\mu,
\end{aligned}
$$

and are characterized by this equation. Hence if we are given T^μ_ν we may compute T and $S(p$ and $\rho(c^2 + \epsilon))$ and verify if Eq. (2.11) is satisfied. If it is, we may then be assured that it is the stress energy tensor of a perfect fluid and determine its velocity field by determining the timelike proper vector of T^μ_ν. From this point of view there is no reason for imposing a restriction such as that given by the inequality (2.6) on ϵ. We shall see, however, that unless an equality such as (2.6) is imposed, we shall not be able to insure that the velocity of sound in the fluid is less than the velocity of light, as it must be in special relativity.

In case the distribution function departs from the Maxwell one, by quantities of the first order, one may show [1]

$$W^{\mu\nu} = -ph^{\mu\nu} - \nu h^{\mu\rho}h^{\nu\sigma}(u_{\rho,\sigma} + u_{\sigma,\rho} - \tfrac{2}{3}u^\lambda_{,\lambda}h_{\rho\sigma}) - \kappa h^{\mu\nu}u^\lambda_{,\lambda}$$

and

$$W^\mu = \lambda h^{\mu\nu}\tilde{\theta}_{,\nu},$$

where p and u are computed from the Maxwell distribution function and

$$\tilde{\theta} = \frac{w + p - TS}{c^2 T}$$

with S the specific entropy, and T the temperature defined below as

$$p = \frac{k}{m}\rho T,$$

k being the Boltzmann constant. λ is the heat conductivity, and ν and κ are the coefficients of viscosity.

3. THE CONSERVATION EQUATIONS

In view of the definitions given above, we may write these five equations (Eqs. (1.8) and (1.10) with $F^\mu = 0$), as

$$(3.1) \qquad (\rho u^\mu)_{,\mu} = 0,$$

$$(3.2) \qquad w_{,\nu}u^\nu + wu^\nu_{,\nu} - W^\mu u_{\mu,\nu}u^\nu + W^\nu_{,\nu} - u_{\mu,\nu}W^{\mu\nu} = 0,$$

and

$$(3.3) \qquad wu^\mu_{,\nu}u^\nu + h^\mu_\lambda(W^{\lambda\nu}_{,\nu} + W^\lambda_{,\nu}u^\nu + W^\lambda u^\nu_{,\nu}) = 0.$$

The last two sets of equations follow from $T^{\lambda\nu}_{,\nu} = 0$ by multiplying by u_λ and h^μ_λ, respectively.

We next turn to the second of these equations, which we may write as

$$(c^2 + \epsilon)(\rho u^\nu)_{,\nu} + \rho\epsilon_{,\nu}u^\nu - W^\mu u_{\mu,\nu}u^\nu + W^\nu_{,\nu} - u_{\mu,\nu}W^{\mu\nu} = 0.$$

Now define a rest temperature T and a specific rest entropy by means of the relation

$$(3.4) \qquad T\,dS = d\epsilon + p\,d\left(\frac{1}{\rho}\right).$$

Then if we make use of the conservation of mass, we have

$$\rho\left(TS_{,\nu} + \frac{p}{\rho^2}\rho_{,\nu}\right)u^\nu - W^\mu u_{\mu,\nu}u^\nu + W^\nu_{,\nu} - u_{\mu,\nu}W^{\mu\nu} = 0$$

or

$$\rho TS_{,\nu}u^\nu - W^\mu u_{\mu,\nu}u^\nu + W^\nu_{,\nu} - u_{\mu,\nu}(W^{\mu\nu} + ph^{\mu\nu}) = 0.$$

This may also be written as

$$(\rho u^\nu S)_{,\nu} + \left(\frac{W^\nu}{T}\right)_{,\nu} + \frac{W^\mu}{T^2}(T_{,\mu} - Tu_{\mu;\nu}u^\nu) + u_{\mu,\nu}P^{\mu\nu} = 0.$$

C. Eckart [2] pointed out that the third and fourth terms in this expression must be negative and hence writes

$$(3.5) \qquad W^\mu = kh^\mu_\lambda[T_{,\mu} - Tu_{\mu,\nu}u^\nu]$$

and

(3.6) $$P^{\mu\nu} = -\lambda h^{\mu\sigma} h^{\nu\tau}[u_{\sigma,\tau} + u_{\tau,\sigma} - \tfrac{2}{3} h_{\sigma\tau} u^{\nu}_{,\nu}].$$

Aside from the bulk viscosity found by Israel, these expressions are consistent with those obtained from the linearized Boltzmann equation.

In the case of a perfect fluid we have

(3.7) $$S_{,\mu} u^{\mu} = 0,$$

as in classical theory.

We note that even if we do not use $(\rho u^{\mu})_{,\mu} = 0$ the equations $T^{\mu\nu}_{,\nu} = 0$ for a perfect fluid may be written as

(3.8) $$w_{,\nu} u^{\nu} + (w + p) u^{\nu}_{,\nu} = 0$$

and

(3.9) $$(w + p) u^{\mu}_{,\nu} u^{\nu} - p_{,\nu} h^{\mu\nu} = 0.$$

We note that if the flow is such that $p = p(w)$ we may define a thermodynamic variable σ such that

(3.10) $$\frac{d\sigma}{\sigma} = \frac{dw}{w + p},$$

and then the first equation becomes

(3.11) $$(\sigma u^{\nu})_{,\nu} = 0$$

and the second becomes

(3.12) $$u^{\mu}_{,\nu} u^{\nu} = -\varphi_{,\nu} h^{\nu\mu},$$

where

(3.13) $$e^{\varphi} = \frac{\sigma}{w + p}.$$

We further note that if the flow is isentropic—that is, if S is the same constant everywhere, not only along streamlines—

$$\sigma = \rho c^2,$$

for in that case

$$d\epsilon = +\frac{p}{\rho^2}\, d\rho,$$

$$dw = d\rho(c^2 + \epsilon) + \rho\, d\epsilon = d\rho(c^2 + \epsilon + p/\rho),$$

$$dw = \frac{d\rho}{\rho} c^2(1 + i/c^2),$$

$$i = \epsilon + p/\rho,$$

and

$$w + p = \rho(c^2 + \epsilon + p/\rho) = \rho c^2(1 + i/c^2).$$

Hence

$$\frac{d\sigma}{\sigma} = \frac{d\rho}{\rho}$$

and

$$e^\varphi = \frac{1}{1 + i/c^2},$$

i being the specific enthalpy.

The derivatives of the velocity vector may be written as

$$u_{\mu,\nu} = \omega_{\mu\nu} + \sigma_{\mu\nu} + \tfrac{1}{3}\theta h_{\mu\nu} + u_{\mu,\nu}u^\rho u_\nu,$$

where

(3.14) $$\omega_{\mu\nu} = \tfrac{1}{2}(u_{\sigma,\tau} - u_{\tau,\sigma})h_\mu^\sigma h_\nu^\tau,$$

(3.15) $$\sigma_{\mu\nu} + \tfrac{1}{3}\theta h_{\mu\nu} = \tfrac{1}{2}(u_{\sigma,\tau} + u_{\tau,\sigma})h_\mu^\sigma h_\nu^\tau,$$

$$g^{\mu\nu}\sigma_{\mu\nu} = h^{\mu\nu}\sigma_{\mu\nu} = 0,$$

(3.16) $$\theta = u_{,\nu}^\nu;$$

thus

(3.17) $$\sigma_{\mu\nu} = \tfrac{1}{2}h_\mu^\sigma h_\nu^\tau[u_{\sigma,\tau} + u_{\tau,\sigma} - \tfrac{2}{3}\theta h_{\sigma\tau}].$$

We shall see later that $\omega_{\mu\nu}$ is related to the vorticity of the flow and that $\sigma_{\mu\nu}$ determines the shear of the flow.

The equations of motion of a perfect fluid, the equations $T_{,\nu}^{\mu\nu} = 0$, then determine

$$\theta = -\frac{w_{,\nu}u^\nu}{w + p} = -\frac{\sigma_{,\nu}}{\sigma} u^\nu,$$

and

$$A^\mu = u_{,\nu}^\mu u^\nu = \frac{p_{,\nu}h^{\nu\mu}}{w + p} = -\varphi_{,\nu}h^{\nu\mu}.$$

In both sets of equations the last equation obtains if $p = p(w)$, in particular when the flow is isentropic.

4. THE RANKINE-HUGONIOT EQUATIONS

From a study of Eqs. (1.11) and (1.12) in Minkowski space-time for the case of one-dimensional motion of a perfect fluid (where $u^2 = u^3 = 0$ and all quantities are functions of x^1 and x^4 alone), it can be shown [3] that the ratio of the velocity of sound to the velocity of light is given by

$$(4.1) \qquad a^2 = \frac{\rho}{(1 + i/c^2)} \left(\frac{di}{d\rho}\right)_S = \left(\frac{\partial p}{\partial w}\right)_S,$$

where the entropy S is kept constant in evaluating $di/d\rho$.

In case the enthalpy is a function of $x = p/\rho c^2$ alone, we may write Eq. (4.1) as

$$a^2 = i'x/(i' - c^2)(1 + i/c^2),$$

where

$$i' = di/dx$$

and

$$\frac{\rho}{c^2} \frac{di}{d\rho} = \frac{\rho i'}{c^2} \frac{dx}{d\rho} = \rho \frac{dx}{d\rho} + x;$$

the last equation is a consequence of the fact that $dS/d\rho = 0$.

A fluid for which the equality holds in (2.6) will be called a limiting fluid. For such a fluid

$$1 + i/c^2 = 5x/2 + (1 + 9x^2/4)^{1/2}$$

and

$$(4.2) \qquad \lim_{x \to \infty} a^2 = \tfrac{1}{3},$$

$$(4.3) \qquad \lim_{x \to \infty} \rho(c^2 + \epsilon) = 3p.$$

This limit for the velocity of sound in an extreme relativistic gas was obtained by Curtis [4] and de Hoffmann and Teller [5]. Taub [3] and Israel [6] have shown by considering sound waves as infinitesimal shock waves the relation between the requirement that the speed of sound be less than the speed of light and certain conditions that must obtain between the various thermodynamic variables that characterize the fluid.

The discussion of one-dimensional motion of a fluid in Minkowski space given in [3] shows that shocks form from compressive motions just as in the classical theory. When this occurs, Eqs. (1.11) and (1.12) cannot be applied at the shocks and must be replaced by statements that relate the variables describing the flow of matter, momentum, and energy across the shocks. These are the relativistic Rankine-Hugoniot equations. They are derived as follows: Eqs. (1.11) and (1.12) are equivalent to the statements that

$$(4.4) \qquad (\rho u^\mu f)_{;\mu} = \rho u^\mu f_{,\mu}$$

and

$$(4.5) \qquad (T^{\mu\nu}\lambda_\nu)_{;\mu} = T^{\mu\nu}\lambda_{\mu;\nu}$$

for arbitrary functions f and vector fields λ_μ which have continuous first derivatives.

In view of Gauss' theorem we may write these equations as

$$(4.6) \qquad \int \rho u^\mu f n_\mu \, dv = \int \rho u^\mu f_{,\mu}(-g)^{1/2} \, d^4x$$

and

$$(4.7) \qquad \int T^{\mu\nu}\lambda_\nu n_\mu \, dv = \int T^{\mu\nu}\lambda_{\mu;\nu}(-g)^{1/2} \, d^4x.$$

The integrals on the right in Eqs. (4.6) and (4.7) refer to an arbitrary four-dimensional volume, and those on the left refer to a closed hypersurface enclosing this volume. Equations (4.6) and (4.7) are meaningful even when the integrands are discontinuous. We assume that they hold in case ρ, u^μ and $T^{\mu\nu}$ are discontinuous across a hypersurface Σ in Minkowski space which is the history of a two-dimensional surface in the spaces $t = $ constant.

By enclosing an arbitrary portion Σ' of the hypersurface in a thin four-dimensional volume and taking the limit as this volume shrinks to zero, we may show that

$$(4.8) \qquad \int_{\Sigma'} [\rho u^\mu] f n_\mu \, dv = 0$$

and

$$(4.9) \qquad \int_{\Sigma'} [T^{\mu\nu}]\lambda_\nu n_\mu \, dv = 0,$$

where we have used the notation

$$[F] = F_+ - F_-,$$

where F_+, F_- are the boundary values of F on the two sides of Σ.

Since the integrals in Eqs. (4.8) and (4.9) refer to arbitrary portions of Σ and since f and λ_μ are arbitrary, we must have

(4.10) $[\rho u^\mu]n_\mu = 0,$

(4.11) $[T^{\mu\nu}]n_\mu = 0.$

In case the $T^{\mu\nu}$ entering into Eqs. (4.11) is given by Eqs. (2.10), Eqs. (4.10) and (4.11) are called the relativistic Rankine-Hugoniot equations, the equations that express the conservation of mass, energy, and momentum across a singular hypersurface in space-time, a hypersurface on which the flow variables may be discontinuous. The derivation given above holds in a general space-time.

If in Minkowski space-time in an inertial coordinate system the parametric equations of Σ are given by

(4.12) $x^\mu = X^\mu(u, v, t),$

that is, if

(4.13)
$$t = t,$$
$$x^i = X^i(u, v, t),$$

then the three-dimensional velocity of a point labelled by u, v on the hypersurface is given by

$$V^i = \partial X^i / \partial t,$$

and the three-dimensional normal to the surfaces in three-space given by the last three of Eqs. (4.13) has components proportional to

$$\Lambda_i = \epsilon_{ijk} \frac{\partial X^j}{\partial u} \frac{\partial X^k}{\partial v}.$$

We normalize the Λ_i by requiring

$$\sum_{i=1}^{3} \Lambda_i^2 = 1.$$

Then the quantity

$$V = \Lambda_i V^i$$

is the velocity of these surfaces in three-space in the direction of the normal. The four-vector

$$n_\mu = \left(\Lambda_1, \Lambda_2, \Lambda_3, \frac{V}{c}\right)$$

is normal to the hypersurface Σ. We may verify that

$$n_\mu n^\mu = \frac{V^2}{c^2} - \sum_i \Lambda_i^2 \leqq (\sum \Lambda_i^2)\left(\frac{\sum V_i^2}{c^2} - 1\right),$$

and hence

$$n_\mu n^\mu \leqq 0$$

is equivalent to

(4.14) $$\sum_{i=1}^{3} V_i^2 \leqq c^2.$$

We define a unit normal vector

$$N_\mu = \alpha n_\mu$$

such that

$$N_\mu N^\mu = -1$$

when Eq. (4.14) holds. Note that if u^μ is the four-velocity of the fluid, then

$$u^\mu = (1 - u^2/c^2)^{-1/2}(u_1/c, u_2/c, u_3/c, 1),$$

where u_i are the components of the three-dimensional velocity of the fluid and

$$u^\mu N_\mu = (1 - V^2/c^2)^{-1/2}(1 - u^2/c^2)^{-1/2}(V - U)/c,$$

where

$$U = \sum \Lambda_i u_i$$

is the three-dimensional velocity of the fluid in the direction of the normal to the surface. Thus $u^\mu N_\mu$ is proportional to the normal component of the velocity of the surface relative to the fluid.

When Eqs. (3.1) hold, Eqs. (4.10) and (4.11) may be written as

(4.15) $$M = \rho_+ u_+^\mu N_\mu = \rho_- u_-^\mu N_\mu$$

and

(4.16) $$M(\mu_+ u_+^\mu - \mu_- u_-^\mu) = (p_+ - p_-)N^\mu$$

with

(4.17) $$\mu = c^2 + i = c^2 + \epsilon + p/\rho,$$

respectively. If V_i are the components of a unit three-vector in the surface, then the four-vector

$$Y^\mu = (V_1, V_2, V_3, 0)$$

satisfies the relation

$$Y^\mu N_\mu = 0,$$

and

$$Y^\mu u_\mu = (1 - u^2/c^2)^{-1/2}(V_i u_i/c)$$

is proportional to the velocity of the fluid in the direction V_i. Multiplying Eqs. (4.16) by Y_μ and summing, we obtain

(4.18) $$M(\mu_+ u_+^\mu Y_\mu - \mu_- u_-^\mu Y_\mu) = 0.$$

Hence, either

(4.19) $$M = 0,$$

or

(4.20) $$\mu_+ u_+^\mu Y_\mu = \mu_- u_-^\mu Y_\mu,$$

or both (4.19) and (4.20) hold. We shall say that the case $M = 0$ represents a slip-stream discontinuity or a density discontinuity and say that the case $M \neq 0$ represents a shock wave. In the former case it follows from Eqs. (4.15) and the interpretation of $u^\mu N_\mu$ that no matter crosses the surface of discontinuity. That is, this surface is made up of streamlines of the fluid. Equation (4.20) need not hold when $M = 0$. If it does in this case, we shall say that there is a density discontinuity present.

If ρ_- and $T_-^{\mu\nu}$ both vanish, we are dealing with a particular case for which $M = 0$. And then Eqs. (4.16) and (4.15) become

$$p_+ = 0$$

and

$$\rho_+ u_+^\mu N_\mu = 0.$$

It is evident that ρ_+ need not vanish if

$$u_+^\mu N_\mu = 0,$$

that is, if the stream-lines of the flow do not cross the surface of discontinuity.

5. SHOCK WAVES

As was stated earlier for such waves, we have $M \neq 0$. It is convenient to rewrite Eqs. (4.15) and (4.16) in terms of the vector

(5.1) $$V_\lambda = (\mu/c^2)u_\lambda$$

and the quantity

(5.2) $$\tau = \mu/\rho.$$

Then

(5.3) $$V^\mu V_\mu = \mu^2/c^4 = (1 + i/c^2)^2,$$

and we have, instead of Eqs. (5.15) and (5.16), the equations

(5.4) $$M = \frac{c^2}{\tau_+} V^\mu_+ N_\mu = \frac{c^2}{\tau_-} V^\mu_- N_\mu,$$

(5.5) $$M(V^\mu_+ - V^\mu_-) = (p_+ - p_-)N^\mu.$$

Equations (4.20), which must hold for shock waves, become

(5.6) $$V^\mu_+ Y_\mu = V^\mu_- Y_\mu$$

for any Y_μ satisfying

$$Y^\mu N_\mu = 0,$$
$$Y^\mu Y_\mu = -1.$$

Equations (5.6) are two of the four equations contained in Eqs. (5.5). The remaining two will be derived below.

Multiply Eqs. (5.5) by $V_{+\mu}$ and by $V_{-\mu}$ and sum and obtain

$$\left(\frac{\mu^2_+}{c^4} - V^\mu_- V_{+\mu} \right) = (p_+ - p_-) \frac{1}{c^2 \tau_+},$$

$$\left(V^\mu_+ V^\mu_- - \frac{\mu^2_-}{c^4} \right) = (p_+ - p_-) \frac{1}{c^2 \tau_-}.$$

Hence

(5.7) $$\frac{\mu^2_+}{c^4} - \frac{\mu^2_-}{c^4} = \frac{1}{c^2} (p_+ - p_-)(\tau_+ - \tau_-).$$

When ϵ is a known function of p and ρ, i is determined as a function of p and ρ and so is τ. Then for fixed p_-, ρ_-, the p_+, ρ_+ satisfying

Eq. (5.7) lie on a curve. This curve connects those (p, ρ)-states that can be obtained from (p_-, ρ_-) by passing through a shock. The curve is known as the Hugoniot curve. We must select that part of the curve passing through p_-, ρ_- corresponding to states p_+, ρ_+ whose entropy $S_+ > S_-$. Otherwise, the second law of thermodynamics would be violated.

Multiply Eqs. (5.5) by N_μ and sum. We obtain

$$(5.8) \qquad \frac{M^2}{c^2}(\tau_+ - \tau_-) = -(p_+ - p_-)(-N_\mu N^\mu)$$

$$= -(p_+ - p_-).$$

Equations (5.4), (5.6), (5.7), and (5.8) are equivalent to Eqs. (5.4) and (5.5).

Since shock waves must travel with a velocity less than that of light and since M must be real, we must have

$$\frac{p_+ - p_-}{\tau_+ - \tau_-} > 0,$$

that is, when

$$p_+ > p_-.$$

In other words, an increase (decrease) in pressure across a shock wave must correspond to a decrease (increase) in τ.

Lichnerowicz [7] has shown that if

$$\left(\frac{\partial \tau}{\partial p}\right)_S < 0,$$

$$\left(\frac{\partial \tau}{\partial s}\right)_p > 0,$$

$$\left(\frac{\partial^2 \tau}{\partial p^2}\right)_S > 0,$$

then $S_+ > S_-$ implies that $p_+ \geq p_-$, $\mu_+ \geq \mu_-$, $\tau_+ \leq \tau_-$ and that the velocity of the shock is less than the velocity of light. The above conditions are generalizations of conditions given by Weyl in his discussion of shocks in prerelativity theory. It may be also shown that if S and p label a point on the Hugoniot curve through S_-, p_- then

$$S - S_- = \frac{1}{12(1 + i/c^2)T} \left(\frac{\partial^2 \tau}{\partial p^2}\right)_s (p - p_-)^3 + \cdots,$$

where T is the temperature.

6. LAGRANGIAN AND CO-MOVING COORDINATES

Classical hydrodynamics deals with a fluid moving in three-space in which a coordinate system is introduced in which every point of the three-space is labelled by coordinates x^i. Relative to this coordinate system, the velocity vector of the fluid is described by its components $v^i(x, t)$. The v^i are called the Eulerian velocities. The streamlines of the fluid are then the curves given by the solutions of the ordinary differential equations

$$(6.1) \qquad dx^i/dt = v^i(x, t), \qquad i = 1, 2, 3.$$

These solutions may be written as

$$(6.2) \qquad x^i = x^i(\xi, t),$$

where the ξ^i are the initial values of the x^i. That is,

$$(6.3) \qquad x^i = x^i(\xi, 0) = \xi^i.$$

The ξ^i are called the Lagrange coordinates and may be used to describe the motion of the fluid. Equations (6.2), supplemented by the equation $t = \tau$, may be regarded as a set of equations describing transformation between the variables t, x^i, and τ, ξ^i, and the equations of conservation of mass, momentum, and energy may be rewritten in terms of the variables τ, ξ^l.

The congruence of curves given by Eqs. (6.2) may be pictured in space-time. In this interpretation, they represent the world lines of an element of the fluid which is labelled by the three parameters ξ^i. These world lines may be described in terms of their four-dimensional tangent vector

$$(6.4) \qquad dx^\mu/dt = w^\mu(x), \qquad \mu = 1, 2, 3, 4,$$

where

$$w^i = v^i, \qquad v^4 = c,$$

since

$$x^4 = ct.$$

We define a hypersurface Σ in space-time by the equation

(6.5)
$$t = 0.$$

In this hypersurface we label points by the coordinates ξ^i. That is, we describe the hypersurface Σ by the parametric equations

(6.6)
$$x^\mu = X^\mu(\xi^1, \xi^2, \xi^3)$$

instead of the functional equation

$$F(x^\mu) \equiv t = 0.$$

Then the congruence of world lines given by Eq. (6.2) may be described as the set of solutions of Eq. (6.4) which intersect the hypersurface Σ in the points ξ^i of Σ.

Equations (6.4) may be written as

$$dx^\mu/dt = \rho u^\mu,$$

where u^μ is the unit four-dimensional velocity vector of the fluid and ρ is given by the equation

$$\rho^2 = \frac{ds^2}{dt^2} = g_{\mu\nu} \frac{dx^\mu}{dt} \frac{dx^\nu}{dt} = g_{\mu\nu} w^\mu w^\nu.$$

If the parameter that labels points on the world line determined by ξ^i is changed to any other parameter, say ω, Eq. (6.4) may still be written as

(6.7)
$$dx^\mu/d\omega = \rho u^\mu = \overline{w}^\mu,$$

where now ω is related to ρ by a similar equation to the one given above. If ω is taken to be equal to s, the proper time along the world line, we have

$$dx^\mu/ds = u^\mu.$$

Now let Σ be an arbitrary spacelike hypersurface in space-time described by the parametric Eqs. (6.6), and let us write the solutions of Eqs. (6.7) as

$$x^\mu = x^\mu(\eta; \omega),$$

where

$$x^\mu = x^\mu(\eta; 0) = \eta^\mu.$$

That is, the η^μ are the initial values of the x^μ. We further require that the η^μ lie on the hypersurface Σ. Then

$$\eta^\mu \equiv X^\mu(\xi),$$

and we may write the solutions of Eqs. (6.7) as

(6.8) $$x^\mu = x^\mu(\xi; \omega).$$

These equations may be regarded as a transformation of co-ordinates from the coordinates x^μ to the coordinates ξ^1, ξ^2, ξ^3, and ω. The latter coordinates are called co-moving coordinates in space-time. They differ from Lagrangian coordinates in that the hypersurface Σ used in their definition need not be given by Eq. (6.5) in the Eulerian coordinate system.

The functions x^μ and $\overline{w}^\mu$ entering into Eqs. (6.7) may be regarded as functions of ξ and ω in view of Eqs. (6.8). If they are so regarded, then they should be written as

(6.9) $$\partial x^\mu / \partial \omega = \overline{w}^\mu(\xi, \omega),$$

and it should be understood that the ξ^i are kept constant in the differentiation involved on the left-hand side of these equations.

From the transformation law of vectors, it follows that in the co-moving coordinate system, the components of the vector $\overline{w}^\mu$ are given by

$$\overline{w}^{\mu*} = \overline{w}^\rho \, \partial x^{\mu*}/\partial x^\rho,$$

where

$$x^{i*} = \xi^i, \qquad x^{4*} = \omega.$$

In view of Eq. (6.9), we have

$$\overline{w}^{\mu*} = \frac{\partial x^{\mu*}}{\partial x^\rho} \frac{\partial x^\rho}{\partial w} = \frac{\partial x^{\mu*}}{\partial x^{4*}} = \delta_4^\mu.$$

Hence, in a co-moving coordinate system, the four-velocity vector of the fluid satisfies the equations

(6.10) $$u^\mu = \delta_4^\mu \, g_{44}^{-1/2},$$

since we must have

$$g_{\mu\nu}u^\mu u^\nu = 1.$$

7. THE EQUATIONS OF HYDRODYNAMICS IN CO-MOVING COORDINATES

Co-moving coordinates may be introduced both in special relativity and in general relativity. In the former theory, they complicate the metric in that we no longer have

$$g_{\mu\nu} = \eta_{\mu\nu},$$

where $-\eta_{11} = -\eta_{22} = -\eta_{33} = \eta_{44} = 1$ and all other $\eta_{\mu\nu} = 0$. However, Eqs. (4.4), (4.6), and (4.7) have relatively simple forms in co-moving coordinates.

Thus if Eq. (6.10) holds, Eq. (3.7) becomes

$$S_{,4} = 0;$$

that is,

(7.1) $$S = S(\xi).$$

Equation (3.1) may be written as

$$((-g)^{1/2}g_{44}^{-1/2}\rho)_{,4} = 0$$

or as

(7.2) $$\rho = g_{44}^{1/2}(-g)^{-1/2}\rho_0(\xi),$$

where g is the determinant of the $g_{\mu\nu}$ and

(7.3) $$g_{\mu\nu} = \eta_{\alpha\beta} \frac{\partial x^\alpha}{\partial x^{*\mu}} \frac{\partial x^\beta}{\partial x^{*\nu}},$$

$$ds^2 = g_{44}\, d\omega^2 + 2g_{4i}\, d\omega\, d\xi^i + g_{ij}\, d\xi^i\, d\xi^j$$

in the case of special relativity. They are such that the Riemann-Christoffel tensor vanishes:

(7.4) $$R^\mu_{\nu\sigma\tau} = 0.$$

Since

$$u^\mu_{;\nu} = \frac{\partial u^\mu}{\partial x^{*\nu}} + u^\sigma \Gamma^\mu_{\sigma\nu}$$

where the $\Gamma^\mu_{\sigma\rho}$ are the Christoffel symbols evaluated in this coordinate system, we have in this coordinate system:

$$u^\mu_{;\nu} = \frac{\partial}{\partial x^{*\nu}}\left(\frac{1}{g_{44}^{1/2}}\right)\delta^\mu_4 + \frac{1}{g_{44}^{1/2}}\Gamma^\mu_{4\nu}$$

and

$$u^\nu u^\mu_{;\nu} = \frac{1}{g_{44}^{1/2}}\frac{\partial}{\partial \omega}\left(\frac{1}{g_{44}^{1/2}}\right)\delta^\mu_4 + \frac{1}{g_{44}}\Gamma^\mu_{44},$$

where

$$\Gamma^\mu_{44} = g^{\mu\nu}\left(\frac{\partial g_{4\nu}}{\partial \omega} - \frac{1}{2}\frac{\partial g_{44}}{\partial x^{*\nu}}\right).$$

If we define

$$(7.5) \qquad \varphi = \log g_{44}^{1/2},$$

we may write

$$u^\nu u_{;\nu}^\mu = g^{\mu\nu} - \frac{\delta_4^\mu \delta_4^\nu}{g_{44}} \frac{\partial \varphi}{\partial x^{*\nu}} + g^{\mu\nu} \frac{\partial}{\partial \omega} \frac{g_{\nu 4}}{g}.$$

Equations (3.9) then become

$$(7.6) \qquad \left(g^{\mu\nu} - \frac{\delta_4^\mu \delta_4^\nu}{g_{44}} \right) \left(\frac{\partial p}{\partial x_*{}^\nu} + \rho c^2 \left(1 + \frac{i}{c^2} \right) \frac{\partial \varphi}{\partial x^{*\nu}} \right)$$

$$= \rho c^2 \left(1 + \frac{i}{c^2} \right) g^{\mu\nu} \frac{\partial}{\partial \omega} \frac{g_{4\nu}}{g_{44}}.$$

It follows from this equation that if the flow is stationary in co-moving coordinates, that is, if all quantities are independent of ω, the time variable in co-moving coordinates, then we must have

$$(7.7) \qquad \frac{\partial p}{\partial \xi^i} + \rho c^2 \left(1 + \frac{i}{c^2} \right) \frac{\partial \varphi}{\partial \xi^i} = 0.$$

If we recall Eq. (7.5) and interpret g_{44} in terms of the Newtonian gravitational potential, then Eq. (7.7) is recognized as the equation for hydrostatic equilibrium.

It is a consequence of Eqs. (7.7) that the pressure is a function of φ, and further that

$$w = \rho(c^2 + \epsilon)$$

is also a function of φ. Hence we have the result that for stationary flows the pressure is a function of the energy density w. Thus the function σ defined by Eq. (3.10) exists for such flows and Eqs. (3.10) and (3.11) obtain. In the co-moving coordinate system the solution to Eq. (3.11) is

$$(7.8) \qquad \sqrt{-g} \, \sigma = e^\varphi f(\xi^i),$$

where $f(\xi^i)$ is a function of the three variables ξ^1, ξ^2, and ξ^3 but not of ω.

When $p = p(w)$, Eq. (7.6) may be written as

$$(7.9) \qquad \frac{\partial}{\partial \xi^i} \left(\frac{w + p}{\sigma} e^\varphi \right) = \frac{\partial}{\partial \omega} \left(\frac{w + p}{\sigma} e^{-\varphi} g_{4i} \right).$$

The integrability conditions of these equations lead to the result that

$$(7.10) \quad \frac{\partial}{\partial \xi^i}\left(\frac{w + p}{\sigma}\, e^{-\varphi} g_{4i}\right) - \frac{\partial}{\partial \xi^i}\left(\frac{w + p}{\sigma}\, e^{-\varphi} g_{4j}\right) = F_{ij}(\xi),$$

where $F_{ij} = -F_{ji}$ are functions of the ξ^i alone and satisfy

$$\frac{\partial F_{ij}}{\partial \xi^k} + \frac{\partial F_{jk}}{\partial \xi^i} + \frac{\partial F_{ki}}{\partial \xi^j} = 0.$$

That is,

$$F_{ij} = \frac{\partial c_i}{\partial \xi^j} - \frac{\partial c_j}{\partial \xi^i},$$

where the c_i are functions of the ξ^i but not of ω.

Hence we must have

$$(7.11) \qquad g_{4i} = \frac{e^\varphi \sigma}{w + p}\left(c_i(\xi) + \frac{\partial \psi}{\partial \xi^i}\right)$$

as the solution of Eqs. (7.10), where ψ is an arbitrary function of the ξ^i and ω. On substituting Eqs. (7.11) into Eqs. (7.9) we find that we must have

$$(7.12) \qquad e^\varphi = \frac{\sigma}{w + p}\left(\frac{\partial \psi}{\partial \omega} + k(\omega)\right),$$

where $k(\omega)$ is an arbitrary function of its argument.

It is no restriction to take

$$\psi = \text{constant},$$

$$k(\omega) = 1,$$

for we can always make a coordinate transformation in space-time which carries the co-moving coordinate system into a co-moving one and in which

$$(7.13) \qquad g_{4i} = \frac{e^\varphi \sigma}{w + p}\, c_i(\xi)$$

and

$$(7.14) \qquad g_{44}^{1/2} = e^\varphi = \frac{\sigma}{w + p}.$$

We also have from Eq. (7.8)

$$(7.15) \qquad \sqrt{-g} = \frac{f}{w + p}.$$

Equations (7.8), (7.13), and (7.14) give the solutions of the equations of motion $T^{\mu\nu}_{;\nu} = 0$ in the co-moving coordinate system in case the fluid obeys the equation of state $p = p(w)$. In a special relativistic problem the functions φ and the remaining components g_{ij} of the metric tensor are determined from the condition that the curvature tensor

$$R^{\mu}_{\nu\sigma\tau} = 0.$$

Once the $g_{\mu\nu}$ are known in the co-moving coordinate system, the transformation of coordinates relating this coordinate system to an inertial one may be determined.

8. THE VORTICITY VECTOR

This four-dimensional vector is defined in a general coordinate system in special relativity by the equations

$$(8.1) \qquad v^{\mu} = (-g)^{-1/2}\epsilon^{\mu\nu\sigma\tau}u_{\sigma;\tau}u_{\nu},$$

where g is the determinant of the metric tensor and $\epsilon^{\mu\nu\sigma\tau}$ ($= \epsilon_{\mu\nu\sigma\tau}$) is the alternating tensor density whose components are equal to plus or minus one, depending on whether $(\mu\nu\sigma\tau)$ is an even or odd permutation of (1234). It is a consequence of the definition of these tensor densities that

$$\epsilon^{\mu\nu\sigma\tau}\epsilon_{\mu\alpha\beta\gamma} = \delta^{\nu}_{\alpha}(\delta^{\sigma}_{\beta}\delta^{\tau}_{\gamma} - \delta^{\tau}_{\beta}\delta^{\sigma}_{\gamma}) + \delta^{\sigma}_{\alpha}(\delta^{\tau}_{\beta}\delta^{\nu}_{\gamma} - \delta^{\nu}_{\beta}\delta^{\tau}_{\gamma}) + \delta^{\tau}_{\alpha}(\delta^{\nu}_{\beta}\delta^{\sigma}_{\gamma} - \delta^{\sigma}_{\beta}\delta^{\nu}_{\gamma})$$

Hence it follows from Eq. (8.1) that

$$(8.2) \qquad \sqrt{-g}\, v^{\mu}\epsilon_{\mu\nu\sigma\tau} = u_{\nu}(u_{\sigma;\tau} - u_{\tau;\sigma}) + u_{\sigma}(u_{\tau;\nu} - u_{\nu;\tau}) + u_{\tau}(u_{\nu;\sigma} - u_{\sigma;\nu})$$

and that

$$(8.3) \quad \tfrac{1}{2}\sqrt{-g}\, v^{\mu}\epsilon_{\mu\nu\sigma\tau}u^{\nu} = \tfrac{1}{2}(u_{\alpha;\beta} - u_{\beta;\alpha})(\delta^{\alpha}_{\sigma} - u^{\alpha}u_{\sigma})(\delta^{\beta}_{\tau} - u^{\beta}u_{\tau}) = \omega_{\sigma\tau},$$

where $\omega_{\sigma\tau}$ is defined by Eq. (3.14).

Flows for which $v^{\mu} = 0$ are said to be irrotational. A necessary and sufficient condition for a flow to be irrotational is $\omega_{\mu\nu} = 0$, as follows from Eqs. (8.2) and (8.3).

In a general co-moving coordinate system we have

$$u_\mu = g_{\mu\nu}u^\nu = e^{-\varphi}g_{\mu 4}.$$

In case an equation of state of the form $p = p(w)$ obtains, we have Eqs. (7.13) and (7.14) holding. Then

$$u_4 = \frac{\sigma}{w + p},$$

$$u_i = \frac{\sigma}{w + p}\, c_i(\xi).$$

It then follows that

$$v^4 = \frac{\sigma^2}{f(w + p)}\, c_i c_{j,k}\epsilon^{ijk},$$

$$v^i = \frac{\sigma^2}{f(w + p)}\, \epsilon^{ijk}c_{j,k},$$

where ϵ^{ijk} is the three-dimensional tensor density.

Hence the necessary and sufficient condition that a flow for which $p = p(w)$ be irrotational is that

$$\frac{\partial c_i}{\partial \xi^j} - \frac{\partial c_j}{\partial \xi^i} = 0.$$

This is, of course, a restriction on the initial conditions in the co-moving coordinate system.

A further discussion of the relativistic equations of motion of a perfect fluid may be found in reference [5], where the concept of circulation is treated and the extension of Bernoulli's theorem and other classical results are obtained.

9. GENERAL RELATIVISTIC HYDRODYNAMICS

Problems in this area are of two sorts: (I) those in which the gravitational field of the fluid can be neglected, but the gravitational field due to some other matter or energy must be taken into account, and (II) those in which the only matter and gravitational fields which are present are the fluid and its own gravitational field.

For problems of type I, the basic equations are Eqs. (1.11),

(1.12), (5.10), and (5.11), but now the covariant differentiation is done with respect to a metric tensor $g^{\mu\nu}$, which is determined by the stress-energy tensor $\theta^{\mu\nu}$, the source of the exterior gravitational field and its potentials, through the equations

$$(9.1) \qquad R^{\mu\nu} - \tfrac{1}{2}g^{\mu\nu}R = -kc^2\theta^{\mu\nu},$$

where $R^{\mu\nu}$ is the Ricci curvature tensor computed from the $g^{\mu\nu}$, R is the scalar curvature, and

$$k = 8\pi G/c^2$$

with G the Newtonian constant of gravitation. Thus for this type of problem the only difference between the special theory of relativity and the general theory is that the underlying space-time is changed from a Minkowski one to a general one whose metric tensor satisfies Eq. (9.1).

In problems of type II Eqs. (9.1) with $\theta^{\mu\nu} = T^{\mu\nu}$ hold. Thus we must determine both $T^{\mu\nu}$ and $g^{\mu\nu}$ so that

$$(9.2) \qquad R^{\mu\nu} - \tfrac{1}{2}g^{\mu\nu}R = -kc^2T^{\mu\nu},$$

and Eqs. (2.1) [or (2.10)], (1.11), (1.12), (4.10), and (4.11) obtain. Equations (1.12) are a consequence of Eqs. (9.2). The former may be considered as a first integral of Eqs. (9.2), and this observation has led McVittie [9] to a method for solving problems in prerelativity hydrodynamics.

Equations (2.10), (1.11), and (9.2) are sometimes said to describe the interior problem for the Einstein field equations. The "solution" of this problem requires the determination of the ten components of the tensor $g^{\mu\nu}$, three independent components of the four-velocity vector u^μ, and the two scalar functions p and ρ. The scalar function ϵ is supposed to be known as a function of p and ρ when the nature of the fluid medium is given. Thus one is required to "solve" the ten equations (9.2) when neither the right-hand side nor the left-hand side of these equations is given. Surprising as it may seem, this can be done at least in a variety of cases. In the following we shall see how this can be done by using the algebraic properties of the $T^{\mu\nu}$ satisfying Eqs. (3.1) or by imposing the conservation of mass [Eq. (1.11)] condition and the condition of isentropy and irrotationality.

By substituting for T^μ_ν from Eqs. (9.2), Eqs. (2.11) and (2.12) become a set of partial differential equations for the $g_{\mu\nu}$ alone. The solutions of these equations may be substituted into Eqs. (9.2) to determine T^μ_ν. The equations for the $g_{\mu\nu}$ obtained by the method described above have been called the consistency equations by McVittie [9]. He dealt with an approximate form of these equations and thus obtained various solutions for problems in prerelativity hydrodynamics. Rainich [10] and Misner and Wheeler [11] used a similar technique for formulating the Einstein-Maxwell field equations for the case where the only stress-energy tensor creating the gravitational field is that due to the electromagnetic field, and no electromagnetic sources are present.

When an equation of state of the form $p = p(w)$ is satisfied by a self-gravitating fluid, we may proceed as in Sec. 7 and introduce a co-moving coordinate system in which Eqs. (6.10), (7.5), (7.13), and (7.14) hold. From the latter equation we may determine p (and hence w and σ) as functions of g_{44}. Hence in this coordinate system $T^{\mu\nu}$ is expressed as a function of the $g_{\mu\nu}$ and Eqs. (9.2) become equations for these dependent variables. The resulting equations become the equations of the Newtonian theory for a self-gravitating fluid if terms of higher order than the first in G, Newton's constant of gravitation, are neglected and if the velocities of the fluid are assumed to be small compared to the special theory of relativity velocity of light.

Chandrasekhar and his coworkers [12] have discussed in detail the differences in the Newtonian and the general relativistic equations describing a self-gravitating fluid by examining the higher-order approximations to the Einstein field equations. The stability of a spherically symmetric distribution of a gaseous mass against radial oscillations has also been treated by Chandrasekhar [13]. He has given a variational principle from which a sufficient condition for instability may be derived. Taub [14] has shown the relation between this variational principle and the second variation problem associated with a variational principle from which the Einstein field equations and the equations of motion for a perfect fluid may be derived.

Although no "general" solution of Eqs. (9.2) is available, many exact and approximate solutions have been discussed. In particular, the problem of the one-dimensional motion of a gas analogous to the problem in classical hydrodynamics of the motion generated by pushing a piston into a tube filled with gas has been treated approximately [15, 16]. The procedure starts with the special relativistic solution to the problem and modifies this solution in accordance with the Einstein field equations. It can be seen that the nonlinearities in the problem reside in the special relativistic approximations and that the gravitational corrections satisfy linear equations. Thus, in the zero approximation shocks occur. It is unlikely that the gravitational corrections satisfying linear equations will remove them. We must, therefore, be prepared in general relativity to treat with and allow for coordinate transformations involving discontinuous derivatives. The metric tensor thus becomes singular [17] on certain hypersurfaces. Although such singularities are associated with coordinate transformations, they are essential ones, since the coordinates involved have intrinsic and physical significance.

It has been pointed out that Eqs. (9.2) may be regarded as nonlinear partial differential equations for the components of the metric tensor $g_{\mu\nu}$ in co-moving coordinates. As such, they must be supplemented with a description of the domain of the independent variables x^μ, that is, a description of the topology of space-time, and with a statement of initial and boundary conditions. As the problem to which these equations apply varies, different topologies appear to be appropriate. The question of what topology is appropriate for a given problem is related to the nature of singularities of metric tensors. Such questions are being actively investigated. A summary of recent results may be found in the paper by R. Geroch [18].

REFERENCES

1. Israel, W., *Jour. Math. Phys.*, **4** (1963), 1163.
2. Eckart, C., *Phys. Rev.*, **58** (1940), 919.
3. Taub, A. H., *Phys. Rev.*, **74** (1948), 328.

4. Curtis, A. R., *Proc. Roy. Soc.*, **A200** (1950), 248.

5. de Hoffmann, F., and E. Teller, *Phys. Rev.*, **80** (1950), 692.

6. Israel, W., *Proc. Roy. Soc.*, **A259**, 129.

7. Lichnerowicz, A., *Relativistic Hydrodynamics and Magnetohydrody-namics.* New York: W. A. Benjamin, 1967, p. 177.

8. Taub, A. H., *Arch. Rational Mech. Anal.*, **3** (1959), 312.

9. McVittie, G. C., *General Relativity and Cosmology.* New York: Wiley, 1956, Chaps. VI and VII.

10. Rainich, G. Y., *Trans. Amer. Math. Soc.*, **27** (1925), 106.

11. Misner, C. W., and J. A. Wheeler, *Ann. Phys.*, **2** (1957), 525.

12. Chandrasekhar, S., et al., *Ap. J.*, **142** (1965), 1488; *Ap. J.*, **153** (1969), 45; *Ap. J.*, **158**, 55; *Ap. J.*, **160** (1970), 153.

13. Chandrasekhar, S., *Ap. J.*, **140** (1964), 417.

14. Taub, A. H., *Comm. Math. Phys.*, **15** (1969), 235.

15. ———, *Phys. Rev.*, **107** (1956), 454.

16. ———, *Phys. Rev.*, **107** (1957), 584.

17. ———, *Ill. Jour. Math.*, **1** (1957), 370.

18. Geroch, R., Varrena Lectures, 1969.

19. A. Lichnerowicz, *Théories relativistes de la gravitation et de l'électro-magnétisme*, Paris, Masson: 1955; and reference 7.

20. Synge, J. L., *The Relativistic Gas.* North-Holland Amsterdam: John Wiley-Interscience, 1957.

21. ———, *Relativity, the General Theory.* North-Holland Amsterdam: John Wiley-Interscience, 1960.

22. Landau, L. D., and E. M. Lifshitz, *The Classical Theory of Fields.* Oxford: Pergamon Press, 1962.

DISPERSIVE WAVES
AND VARIATIONAL PRINCIPLES

G. B. Whitham

1. INTRODUCTION

There are two main classes of wave motion. The first might be called "hyperbolic," since the governing partial differential equations are hyperbolic, and the second might be called "dispersive," since in the simplest examples a local disturbance disperses into a wave train. In general, dispersive waves are not governed by hyperbolic equations, and the elucidation of their mathematical structure and typical properties is a more complicated affair.

The distinction between the two classes is not always sufficiently well recognized. Students sometimes are left with the impression that the mathematical theory of hyperbolic equations is a sophisticated approach to wave motion and covers all cases, whereas the use of sinusoidal solutions

$$e^{i(\kappa x - \omega t)},$$

with or without Fourier superposition, is an elementary approach to the same thing. The confusion is possible because both approaches may be used in an attack on the simple wave equation. However, the two approaches are really focussing on the two different classes of wave motion. Furthermore, while many wave problems are hyperbolic, one could probably safely claim that the majority are not. In the most familiar example of water waves, the governing equation is Laplace's equation with strange boundary conditions at the free surface.

The theory of hyperbolic equations, with its attendant discussion of characteristics, shock formation in nonlinear problems, weak solutions, etc., is well established. Even for linear dispersive problems, the corresponding questions of the general form of the equations and the general features of their solutions are much more complicated. For nonlinear dispersive problems such results were almost nonexistent. In the last few years, however, there has been considerable development in the discussion of dispersive waves, both linear and nonlinear. It has led to partial unification in points of view and in techniques. This article is a review of some of these developments with emphasis on those which contribute to the unification and which allow extension to nonlinear problems. A mathematical novelty is the description of waves by variational principles.

2. DISPERSIVE WAVES

The study of dispersive waves starts with periodic traveling wave trains. In linear problems, these are sinusoidal and may be taken in the complex form

$$(2.1) \qquad u(x, t) = ae^{i(\kappa x - \omega t)},$$

where a, κ, ω are real constants. Here, a is the amplitude, κ is the wave number (number of oscillations per 2π in space), and ω is the frequency (number of oscillations per 2π in time). The velocity of the wave train is the phase velocity $c = \omega/\kappa$. The frequency ω and wave number κ are related by a "dispersion relation"

$$(2.2) \qquad G(\omega, \kappa) = 0,$$

where the function G is determined by the particular equations of the problem. Some examples are

Beam equation

$$u_{tt} + \gamma^2 u_{xxxx} \equiv 0,$$

(2.3)

$$G(\omega, \kappa) \equiv \omega^2 - \gamma^2 \kappa^4 = 0,$$

Linear Korteweg-de Vries equation:

$$u_t + \alpha u_x + \beta u_{xxx} = 0,$$

(2.4)

$$G(\omega, \kappa) \equiv \omega - \alpha\kappa + \beta\kappa^3 = 0.$$

Neither of these equations is hyperbolic. Indeed, integral equations may also admit wave train solutions. The equations

(2.5) $$u_t + \int_{-\infty}^{\infty} K(x - \xi)u_\xi(\xi, t)\, d\xi = 0,$$

(2.6) $$G(\omega, \kappa) \equiv \omega - \kappa \int_{-\infty}^{\infty} K(\zeta)e^{-i\kappa\zeta}\, d\zeta = 0,$$

and

(2.7) $$u_{tt} - \int_{-\infty}^{\infty} K_1(x - \xi)u_{\xi\xi}(\xi, t)\, d\xi = 0,$$

(2.8) $$G(\omega, \kappa) \equiv \omega^2 - \kappa^2 \int_{-\infty}^{\infty} K_1(\zeta)e^{-i\kappa\zeta}\, d\zeta = 0$$

are examples. It should be noted in (2.6) that the phase velocity $c(\kappa)$ is the Fourier transform of the kernel $K(x)$, and conversely the kernel $K(x)$ is the Fourier transform

(2.9) $$K(x) = \frac{1}{2\pi} \int_{-\infty}^{\infty} c(\kappa)e^{i\kappa x}\, d\kappa.$$

Similarly, the kernel $K_1(x)$ in (2.7) is the transform of $c^2(\kappa)$. These are useful in proposing equations that will provide a *given* dispersion relation; the appropriate kernels K or K_1 are determined from the given $c(\kappa)$ and used in (2.5) or (2.7). When $c(\kappa)$ is a polynomial as in (2.3) and (2.4), the kernels K, K_1 consist of δ-functions, and (2.5) and (2.7) reduce to the corresponding differential equations.

In the linear theory of water waves, there are elementary solutions which lead to the expression (2.1) for the height of the surface, with the dispersion relation

(2.10) $$G(\omega, \kappa) \equiv \omega^2 - g\kappa \tanh \kappa h = 0,$$

where h is the undisturbed depth and g is the acceleration due to gravity. The analysis involves the vertical coordinate y as well as x and t, but the dependence on y is not wavelike. The linear problem is formulated [1] in terms of a velocity potential $\varphi(x, y, t)$ which has to satisfy

$$\varphi_{xx} + \varphi_{yy} = 0, \qquad -h < y < 0,$$

(2.11) $\qquad \varphi_{tt} + g\varphi_y = 0 \qquad$ on $y = 0,$

$$\varphi_y = 0 \qquad \text{on } y = -h.$$

The surface elevation is then given by

$$(2.12) \qquad \eta(x, t) = -g^{-1}\varphi_t(x, 0, t).$$

The periodic wave train solution is

$$(2.13) \qquad \varphi = \frac{ga}{i\omega} \frac{\cosh \kappa(h + y)}{\cosh \kappa h} e^{i(\kappa x - \omega t)}, \qquad \eta = ae^{i(\kappa x - \omega t)},$$

with ω, κ related as in (2.10).

This variety in the equations governing the wave motion shows that the unifying feature is the dispersion relation itself rather than the type of equation. For these linear problems one might almost view the equations as a mere source of the appropriate dispersion relation and push them into a subsidiary role. As noted above, an equation can always be constructed to match a dispersion relation. For simple linear partial differential equations of the form

$$P\left(\frac{\partial}{\partial t}, \frac{\partial}{\partial x}\right) u = 0,$$

as in (2.3) and (2.4), the dispersion relation is just

$$P(-i\omega, i\kappa) = 0,$$

with an immediate correspondence

$$\frac{\partial}{\partial t} \leftrightarrow -i\omega, \qquad \frac{\partial}{\partial x} \leftrightarrow i\kappa$$

between the two. It is clear that such differential equations have a polynomial for P, and hence will always yield polynomial dispersion relations. Equation (2.10) is transcendental because of the nonwavelike dependence on y. The formulation of (2.5) from

(2.9) and (2.10) has been pursued to some extent in [2] and [3].

The extension to more space dimensions is immediate with a vector wave number κ in (2.1) and (2.2) to give

(2.14) $\qquad u = ae^{i(\kappa \cdot x - \omega t)}, \qquad G(\omega, \kappa) = 0.$

The phase velocity has magnitude $c = \omega/|\kappa|$ and direction κ.

For linear problems, more general solutions are constructed by Fourier superposition of solutions (2.1) or (2.14) for different wave numbers κ with corresponding frequencies $\omega(\kappa)$ to satisfy the dispersion relation. The term "dispersion" refers to those cases for which the phase velocity is not the same for all values of κ. For, in such cases, the Fourier components making up any given distribution will travel at different speeds and the disturbance will "disperse." It is convenient to express the condition for dispersive waves in terms of the roots $\omega(\kappa)$ of the dispersion relation, and to change the requirement slightly. We shall refer to linear waves as dispersive if

(2.15) $\qquad \text{determinant} \left| \dfrac{\partial^2 \omega}{\partial \kappa_i \partial \kappa_j} \right| \neq 0.$

For one-dimensional waves this reduces to

(2.16) $\qquad \omega''(\kappa) \neq 0.$

Of course, these quantities might vanish for isolated values of κ, or in the limits of $\kappa \to 0$ and $\kappa \to \infty$. Equation (2.16) is almost equivalent to $c'(\kappa) \neq 0$; it excludes $\omega = \alpha\kappa + \beta$, whereas the latter does not. It turns out that this special example should be excluded because there is no real dispersion. It should also be noted that the simple wave equation with $\omega = \pm\alpha\kappa$ has been eliminated from dispersive waves. The approach via (2.1) and Fourier superposition gives the correct results, but there is no dispersion. The behavior of the genuine cases with $\omega''(\kappa) \neq 0$ is given in the next section.

The dispersion relation, its use in Fourier synthesis, and, as we shall see later, its use in direct asymptotic techniques allow a general treatment of linear problems. But this is clearly insufficient for nonlinear problems. We can identify dispersive wave problems by the existence of periodic wave trains analogous to

(2.1), but Fourier superposition cannot be used for further developments in the nonlinear case.

Probably the first nonlinear dispersive waves were discussed by Stokes in 1847 in his investigations of water waves. He found solutions to the nonlinear counterpart of (2.11) by expanding all quantities in powers of the amplitude, with due care to keep the expansions uniformly valid by allowing the frequency to depend on the amplitude. (The latter technique subsequently became known as Poincaré's method!) For deep water, the expansions for the surface elevation and the frequency are

$$\eta = a \cos (\kappa x - \omega t) + \tfrac{1}{2}\kappa a^2 \cos 2(\kappa x - \omega t)$$
$$+ \tfrac{3}{8}\kappa^2 a^3 \cos 3(\kappa x - \omega t) + \cdots$$

(2.17)

$$\omega^2 = g\kappa + g\kappa^3 a^2 + \cdots.$$

Stokes's investigation introduces the key idea that the dispersion relation includes the amplitude; the phase velocity depends on both wave number and amplitude. Simpler examples in which the results corresponding to (2.17) are expressed in closed form appeared later. One of the most important ones, again from water waves, is the Korteweg-de Vries equation, which is an approximation for water waves in shallow water. It has the advantage that the vertical coordinate has been eliminated, and a single equation is written for the surface elevation. It is

(2.18) $$\eta_t + (c_0 + \alpha\eta)\eta_x + \beta\eta_{xxx} = 0,$$

where $c_0 = \sqrt{gh}$, $\alpha = 3c_0/2h$, $\beta = c_0 h^2/6$. Uniform wave trains are obtained from solutions

(2.19) $$\eta = \eta(\theta), \qquad \theta = \kappa x - \omega t.$$

When this form is substituted in (2.18), the resulting ordinary differential equation can be integrated twice to

(2.20) $$\frac{1}{2}\beta\kappa^2 \left(\frac{d\eta}{d\theta}\right)^2 = A + B\eta + \frac{1}{2}\left(\frac{\omega}{\kappa} - c_0\right)\eta^2 - \frac{1}{6}\alpha\eta^3,$$

where A, B are constants of integration. The solution may then be written in terms of the Jacobian elliptic function $cn\theta$ and, as a consequence, Korteweg and de Vries named these waves "cnoidal waves." A limiting case (when the cubic on the right of (2.20) has a double zero) gives the famous solitary wave. In the periodic case, the relations of the amplitude and modulus of the elliptic function to the parameters ω, κ provides the dispersion relation

$$(2.21) \qquad G(\omega, \kappa, a) = 0.$$

The general conclusion is that nonlinear dispersive waves are recognized by the existence of periodic wave trains

$$(2.22) \qquad u = U(\theta), \qquad \theta = \kappa x - \omega t,$$

where U is a periodic function of the phase θ; the solution for U will bring in a parameter a, which is the amplitude, and the solution requires a dispersion relation in the form (2.21). It turns out that there may be other important parameters that appear both in U and G.

A simpler example for some purposes, which has acquired recent interest, is a nonlinear version of the Klein-Gordon equation:

$$(2.23) \qquad u_{tt} - u_{xx} + V'(u) = 0.$$

This happens to be hyperbolic, but we shall be concerned with dispersive behavior away from wavefronts. (The appearance of both hyperbolic and dispersive behavior in the same equation shows yet again the complications in the classification problem.) The periodic solution satisfies

$$(2.24) \qquad (\omega^2 - \kappa^2)U_{\theta\theta} + V'(U) = 0,$$

which has the first integral

$$(2.25) \qquad \tfrac{1}{2}(\omega^2 - \kappa^2)U_\theta^2 + V(U) = A.$$

The constant A can be used as a parameter equivalent to the amplitude a. Equation (2.25) can be solved to give the function $U(\theta)$ in the inverse form

$$(2.26) \qquad \theta = \sqrt{\tfrac{1}{2}(\omega^2 - \kappa^2)} \int \frac{dU}{\sqrt{A - V(U)}}.$$

We may normalize the period in θ to be 2π, so that κ and ω still determine the number of oscillations per 2π units of length and time. Then, it follows that

$$(2.27) \qquad 2\pi = \sqrt{\tfrac{1}{2}(\omega^2 - \kappa^2)} \oint \frac{dU}{\sqrt{A - V(U)}},$$

where $\oint$ denotes integration over a complete period. This is the dispersion relation between ω, κ and A.

The next step is to see how more general solutions can be developed from these periodic wave trains. In linear problems, the Fourier synthesis is clear. This is taken up next before we return to the nonlinear problems.

3. FOURIER SYNTHESIS AND ASYMPTOTIC BEHAVIOR

Formally, a general solution

$$(3.1) \qquad u = \int_{-\infty}^{\infty} F(\kappa)e^{i\{\kappa x - W(\kappa)t\}} \, d\kappa$$

is deduced from (2.1) and (2.2), where $\omega = W(\kappa)$ is a specific solution of the dispersion relation (2.2). The complete solution will be the sum of terms like (3.1) with one integral for each of the solutions $\omega = W(\kappa)$ of (2.2). The arbitrary functions $F(\kappa)$ would be determined from appropriate initial or boundary conditions. These Fourier integrals give exact solutions, but their content is hard to see, and we are looking for ways to avoid them for later extension to nonlinear problems. One starts to understand the main features of dispersive waves by considering the asymptotic behavior for large x and t. For wave motions, we are interested in traveling disturbances, which means that we are concerned with dependence on $x - Ct$ for various C. Accordingly, we consider the asymptotic behavior of (3.1) as $t \to \infty$, with x/t held fixed. The solution (3.1) is written

$$(3.2) \qquad u = \int_{-\infty}^{\infty} F(\kappa)e^{-i\chi t} \, d\kappa,$$

where

$$(3.3) \qquad \chi(\kappa) = W(\kappa) - \kappa \frac{x}{t}.$$

For the present, x/t is a fixed parameter, and only the dependence on κ is displayed in χ. The integral in (3.2) may then be studied by the method of stationary phase; this is, in fact, the problem for which Kelvin developed the method. Kelvin argued that for large t, the main contribution to the integral is from the neighborhood of stationary points $\kappa = k$ defined by

$$(3.4) \qquad \chi'(k) = W'(k) - \frac{x}{t} = 0.$$

Otherwise, the contributions oscillate rapidly and make little net contribution. The method and the related method of steepest descents, which is easier to use for the full asymptotic expansion, is now standard. (See [4], for example.) The first term in the asymptotic behavior is

$$(3.5) \qquad \begin{aligned} u(x, t) &\sim F(k) \sqrt{\frac{2\pi}{t|\chi''(k)|}}\, e^{-i\chi(k)t - i(\pi/4)\mathrm{sgn}\chi''(k)} \\ &= F(k) \sqrt{\frac{2\pi}{t|W''(k)|}}\, e^{ikx - iW(k)t - i(\pi/4)\mathrm{sgn}W''(k)}. \end{aligned}$$

In this formula k is a function $k(x, t)$ determined by (3.4).

The remarkable significance of (3.5) derives from the fact that it may be written just as in (2.1) as

$$(3.6) \qquad u \sim a e^{i\theta}$$

with the important difference that a is no longer constant, nor are θ_x and θ_t. Indeed, with

$$\theta = kx - W(k)t,$$

we have, from (3.4),

$$(3.7) \qquad \theta_x = k + \{x - W'(k)t\}k_x = k(x, t)$$

$$(3.8) \qquad -\theta_t = W(k) - \{x - W'(k)t\}k_t = W(k).$$

Thus θ_x and $-\theta_t$ still have the significance of a wave number and a frequency, but they are no longer constant. At any point (x, t), they give the "local" wave number and "local" frequency. They are still related by the dispersion relation, and their variations with x and t are determined by (3.4). A vivid interpretation of (3.4) is provided by asking: How should an observer move if he wishes to follow a particular value k_0 of the local wave number? The answer is that he must keep

$$x = W'(k_0)t.$$

That is, he should move with a constant velocity $W'(k_0)$; this is the "group velocity." We may embody this important result in the statement that values of the local wave number *propagate* with the group velocity. In contrast, to keep the same phase θ_0, such as a particular crest, an observer must travel with the phase velocity $W(k)/k$, where k is determined such that

$$\theta_0 = kx - W(k)t,$$

and this phase velocity is *not* constant. The distinction between group and phase velocities is crucial, and the former plays the dominant role.

The complex amplitude in (3.6) is

$$(3.9) \qquad a = F(k) \sqrt{\frac{2\pi}{t|W''(k)|}}\, e^{-i(\pi/4)\mathrm{sgn}W''(k)}.$$

The quantity $|a|^2$ is related to the energy density, and the significant form of (3.9) is related to the conservation of this quantity. Consider

$$(3.10) \qquad Q(t) = \int_{x_1}^{x_2} |a|^2\, dx$$

$$= 2\pi \int_{x_1}^{x_2} \frac{F(k)F^*(k)}{t|W''(k)|}\, dx.$$

In this integral, $k(x, t)$ is given by (3.4). Since k appears in the arguments of the integrand and x does not, it is natural to introduce k as a new variable of integration through the transformation

$$x = W'(k)t.$$

For $W''(k) > 0$, we have

$$Q(t) = 2\pi \int_{k_1}^{k_2} F(k)F^*(k)\, dk,$$

where k_1 and k_2 are defined by

$$(3.11) \qquad x_1 = W'(k_1)t, \qquad x_2 = W'(k_2)t.$$

If $W''(k) < 0$, the limits are reversed. Now, if k_1 and k_2 are held fixed as t varies, $Q(t)$ remains constant. According to (3.11), the points x_1 and x_2 are then moving with the corresponding group velocities. We have shown, therefore, that the total amount of

$|a|^2$ between points moving with the group velocity remains the same. *In this sense, "energy" propagates with the group velocity.* The points (3.11) separate with a distance proportional to t; hence, $|a|$ decrease like $t^{-1/2}$. The relation between $|a|^2$ and the true energy density will appear later.

It is striking that these important results depend only on the dispersion relation $\omega = W(k)$, and particularly on the expression $W'(k)$ for the group velocity. One then has the feeling that the Fourier analysis must be largely irrelevant, and one expects that similar results are feasible for nonlinear problems. We now turn to these questions.

The motivating step is to realize that in these asymptotic results we are dealing with "slowly varying" wave trains. The asymptotic solution (3.6) is the same expression as in the uniform wave train (2.1), but the parameters a, $k = \theta_x$, $\omega = -\theta_t$ are no longer constant. They are, however, slowly varying in the sense that the relative change in one wavelength or one period is small. This is easily seen from (3.4) and (3.9). From (3.4),

$$\frac{k_t}{k} = -\frac{W'(k)}{kW''(k)} \frac{1}{t}, \qquad \frac{k_x}{k} = \frac{1}{kW''(k)} \frac{1}{t},$$

with similar results for ω, and from (3.9),

$$\frac{a_t}{a} = \left\{ \frac{F'(k)}{F(k)} - \frac{W'''(k)}{2W''(k)} \right\} \frac{k_t}{k} - \frac{1}{2t},$$

with a similar expression for a_x/a. Since the asymptotic expansion is for $t \to \infty$ and $x/t = W'(k)$ fixed, these are small.

With this motivation, we consider slowly varying wave trains in a more general context. The idea is to start with the uniform wave train, whether linear or nonlinear, extend the solution to allow the parameters (ω, k, a) and any others that there may be to be slowly varying, and find direct ways of obtaining equations for these slowly varying parameters. By "slowly varying" we mean that their relative changes in one wavelength and in one period are small. The slow variations may come about in other ways than in the asymptotic behavior discussed above. It is of interest to take a signalling problem from a periodic source that is slowly modulated, or to consider propagation into a nonhomogeneous medium whose properties vary slowly in space or time. In these cases the small parameter ϵ for the slow variations will be provided in the boundary conditions or the equations, whereas it was a typical period divided by t itself in the case studied above.

We shall proceed intuitively at first in order to propose methods and derive results. The justification by formal perturbation expansions will be explained in Sec. 9.

4. SIMPLE DERIVATION OF GROUP VELOCITY CONCEPTS FOR LINEAR PROBLEMS

We start with linear problems. We know from Sec. 3 that the group velocity plays two distinct roles. In one it determines the propagation of wave number and frequency; in the other, it determines variations in amplitude. In the asymptotic solution (3.6), one role is concerned with the determination of θ, the other with the determination of a.

To describe a slowly varying wave train there must be a phase function $\theta(x, t)$. A local wave number $k(x, t)$ and frequency $\omega(x, t)$ are defined in terms of θ by

$$(4.1) \qquad k(x, t) = \frac{\partial \theta}{\partial x}, \qquad \omega(x, t) = -\frac{\partial \theta}{\partial t}.$$

In view of the assumed slow variations of k, ω, it is reasonable to propose that these quantities still satisfy the dispersion relation. For linear problems the dispersion relation is

$$(4.2) \qquad G(\omega, k) = 0.$$

Equations (4.1) and (4.2) give a nonlinear partial differential equation for θ. It is convenient to determine k and ω from (4.2) and the consistency relation (elimination of θ from (4.1))

$$(4.3) \qquad \frac{\partial k}{\partial t} + \frac{\partial \omega}{\partial x} = 0.$$

If $\omega = W(k)$ is a solution of (4.2), then (4.3) gives

$$(4.4) \qquad \frac{\partial k}{\partial t} + C(k) \frac{\partial k}{\partial x} = 0, \qquad C(k) = W'(k).$$

This shows immediately that k is constant on characteristic curves defined by

$$\frac{dx}{dt} = C(k)$$

in the (x, t) plane. Since k is constant on each such curve, the curves are straight lines, each with slope $C(k)$ corresponding to the value of k on that curve. The solution for an initial distribution $k = f(x)$ at $t = 0$ is

$$(4.5) \qquad k = f(\xi), \qquad x = \xi + C(k)t.$$

In more physical terms, values of k propagate with the corresponding value of the group velocity $C(k)$. When the disturbance is concentrated initially at the origin, k may be determined from

$$(4.6) \qquad x = C(k)t.$$

This is the case in (3.4); viewed from large x and t, the initial disturbance *is* effectively concentrated at the origin. In this case, but only in this case,

$$\theta = kx - W(k)t.$$

We thus have a very simple and yet more general derivation of the kinematics of the waves.

The second role of the group velocity is in the determination of a, and this involves dynamics. One looks for a differential equation for a to match the one for k in (4.4). It is not hard to find by various routes. Leaning first on (3.10) and (3.11), we note that

$$\frac{dQ}{dt} = \frac{d}{dt} \int_{x_1}^{x_2} |a|^2 \, dx = \int_{x_1}^{x_2} \frac{\partial}{\partial t} |a|^2 \, dx + |a|_2^2 C(k_2) - |a|_1^2 C(k_1) = 0.$$

As $x_2 - x_1 \to 0$, we have

$$(4.7) \qquad \frac{\partial}{\partial t} |a|^2 + \frac{\partial}{\partial x} \{C(k)|a|^2\} = 0$$

in the limit. Alternatively, it is well-known in specific problems that the energy density and energy flux are proportional to $|a|^2$ and $C(k)|a|^2$, respectively. This is, in fact, another popular way to argue that energy propagates with the group velocity. Equa-

tion (4.7) is the corresponding equation for the conservation of
energy. There are many variants of this argument that may be
used to derive (4.7). But, until recently, they all suffered from
the deficiency that (4.7) had to be established separately from
the governing equations for each specific problem. Why is the
answer always the same in terms of the group velocity? This has
now been resolved by a general approach using variational principles. At the same time, it is shown that (4.7) is not basically
the equation for energy conservation, but rather the conservation
of "wave action," which in simple cases is proportional to the
energy.

For nonlinear problems, the introduction of θ and $k = \theta_x$,
$\omega = -\theta_t$, is still sound. But the dispersion relation is now

$$G(\omega, k, a) = 0$$

and involves a. Thus, the two equations corresponding to (4.3)
and (4.7) become coupled. The variational approach has also led
to a simple derivation of the corresponding equations. Before we
take this up in Sec. 6, the linear kinematics is extended to more
dimensions, which could also be done by stationary phase on
multiple Fourier integrals analogous to Sec. 3, and to nonhom-
ogenous media, which cannot be studied by Fourier integrals.
A few examples are also given to show the power of these simple
methods, and to show the value of this type of more intuitive
discussion even when the exact solution in terms of Fourier
integrals is known.

5. EXTENSIONS AND EXAMPLES

In more dimensions, the phase $\theta(\mathbf{x}, t)$ is a function of the vector
position $\mathbf{x}$ and the time t. The vector wave number $\mathbf{k}$ and frequency ω are defined by

$$(5.1) \qquad\qquad k_i = \frac{\partial \theta}{\partial x_i}, \qquad \omega = -\frac{\partial \theta}{\partial t}.$$

For linear problems in a homogeneous medium, the dispersion
relation is $G(\omega, \mathbf{k}) = 0$. This relation will involve the parameters
of the medium. For a nonhomogeneous medium, these parameters

will be slowly varying functions of $\mathbf{x}$ or t, and the dispersion relation becomes

$$(5.2) \qquad G(\omega, \mathbf{k}; t, \mathbf{x}) = 0.$$

We still propose to use it in conjunction with (5.1). The cross-elimination of θ from (5.1) gives

$$(5.3) \qquad \frac{\partial k_i}{\partial t} + \frac{\partial \omega}{\partial x_i} = 0, \qquad \frac{\partial k_i}{\partial x_j} - \frac{\partial k_j}{\partial x_i} = 0.$$

If $\omega = W(\mathbf{k}; t, \mathbf{x})$ is taken as a solution of (5.2), the first equation in (5.3) gives

$$(5.4) \qquad \frac{\partial k_i}{\partial t} + \frac{\partial W}{\partial k_j}\frac{\partial k_j}{\partial x_i} = -\frac{\partial W}{\partial x_i}.$$

We now introduce the vector group velocity defined by

$$(5.5) \qquad C_j = \frac{\partial W}{\partial k_j}$$

and note from the second relation in (5.3) that $\partial k_j/\partial x_i$ may be replaced by $\partial k_i/\partial x_j$. Then

$$(5.6) \qquad \frac{\partial k_i}{\partial t} + C_j \frac{\partial k_i}{\partial x_j} = -\frac{\partial W}{\partial x_i}.$$

In characteristic form, the equations may be written

$$(5.7) \qquad \frac{dk_i}{dt} = -\frac{\partial W}{\partial x_i}, \qquad \frac{dx_i}{dt} = \frac{\partial W}{\partial k_i} = C_i.$$

These are Hamilton's equations with the k_i in the role of the generalized momenta and the frequency W in the role of the Hamiltonian. This is the well-known duality exploited in quantum theory. The equation for θ is

$$\frac{\partial \theta}{\partial t} + W\left(\frac{\partial \theta}{\partial x_i}, t, x_i\right) = 0;$$

this is the Hamilton-Jacobi equation.

In homogeneous media the wave number $\mathbf{k}$ still propagates unchanged with the group velocity $\mathbf{C}(\mathbf{k})$. In nonhomogeneous media, it propagates with this velocity, but the values vary at the rate $-\partial W/\partial \mathbf{x}$. In homogeneous media the general solution corresponding to (5.1) is

$$k_i = f_i(\xi),$$
$$x_i = \xi_i + C_i(\mathbf{k})t.$$

For a concentrated initial distribution, $\mathbf{k}$ is determined by

(5.8) $$x_i = C_i(\mathbf{k})t.$$

(This would be the stationary point corresponding to (3.4) in the multiple Fourier analysis.)

A few typical and interesting examples are taken from the theory of water waves.

OCEAN SWELL FROM STORMS. In deep water, the dispersion relation (2.10) may be approximated by its limiting form $\omega = \sqrt{gk}$, for $kh \to \infty$. For two dimensions this becomes

(5.9) $$\omega = W(\mathbf{k}) = \sqrt{gk}, \qquad k = |\mathbf{k}|.$$

For the wave distribution at large distances from a concentrated storm it is appropriate to use (5.8). The pattern is symmetric, since $W(\mathbf{k})$ depends only on $|\mathbf{k}|$ and not on its direction. The group velocity is

$$C_i = \frac{1}{2}\sqrt{\frac{g}{k}}\frac{k_i}{k}, \qquad C = |\mathbf{C}| = \frac{1}{2}\sqrt{\frac{g}{k}}.$$

Hence, from (5.8),

$$r = |\mathbf{x}| = \frac{1}{2}\sqrt{\frac{g}{k}}\, t.$$

Therefore,

$$k = \frac{1}{4}\frac{gt^2}{r^2}, \qquad \omega = \frac{1}{2}\frac{gt}{r}.$$

The formula has been applied to waves produced by storms in the South Pacific [5]. At distances of the order of 2000 miles, the frequency was found to vary linearly with t, and the constant of proportionality gave a very accurate determination of the distance of the storm.

SHIP WAVES. The wave pattern produced by a ship (or other moving object) traveling with uniform velocity U on the surface of deep water is most easily treated as a steady flow problem relative to the ship. Then $\omega = 0$, but the two components (k_1, k_2)

of the wave number are still related by a dispersion relation, and the geometry is determined according to (5.3) from

(5.10)
$$\frac{\partial k_1}{\partial x_2} - \frac{\partial k_2}{\partial x_1} = 0.$$

The dispersion relation is obtained by a transformation of (5.9) to a moving frame; it becomes

(5.11) $\qquad Uk_1 + \sqrt{gk} = 0, \qquad k = (k_1^2 + k_2^2)^{1/2}.$

The characteristic form of (5.10) shows that the various values of k_1 and k_2 will be distributed on lines

(5.12)
$$\frac{dx_1}{dx_2} = -\frac{dk_2}{dk_1},$$

where $k_2 = f(k_1)$ is determined from (5.11). It is easily verified that the right-hand side of (5.12) has a minimum value of $2\sqrt{2}$ for all values of k_1. Hence, the wave pattern is contained in a wedge-shaped region spreading out behind the ship, and the semi-angle of the wedge is $\cotan^{-1} 2\sqrt{2} = 19.5°$. This is Kelvin's famous result. Full details of the wave pattern can also be found; the version following this approach is given in [6].

WAVES IN A STRATIFIED FLUID. In the so-called Bonssinesq approximation, the dispersion relation is

$$\omega^2 = \frac{\omega_0^2 k_1^2}{(k_1^2 + k_2^2)},$$

where the density ρ_0 is stratified in the x_2 direction and

$$\omega_0^2 = -\frac{g}{\rho_0} \frac{d\rho_0}{dx_2}.$$

First, it is clear that waves are possible only if $\omega < \omega_0$. Second, the group velocity,

$$\mathbf{C} = \left(\frac{\omega_0 k_2^2}{k^3}, -\frac{\omega_0 k_1 k_2}{k^3} \right),$$

is perpendicular to the wave number $\mathbf{k}$. Hence, the group velocity is perpendicular to the phase velocity; this is a striking example of the distinction between the two. Third, if waves are stimulated by a vibrating source with a fixed ω, an analysis similar to the ship wave case shows that the waves are confined to a wedge-shaped region, making an angle $\sin^{-1}(\omega/\omega_0)$ with the horizontal. Full details and impressive photographs of experiments showing all these features are given in [7].

6. VARIATIONAL PRINCIPLES

We return now to the general discussion and continue the development from Sec. 4. As noted there, the appropriate approach is through variational principles. The power of the method appears immediately, since it is just as easy to do nonlinear problems as linear ones. The nonlinear Klein-Gordon equation (2.23) will be used to develop the ideas. Ultimately, all that will be used is its corresponding Lagrangian, so the results can be taken over in general to any Lagrangian.

Equation (2.23) may be derived from the variational principle

$$(6.1) \qquad \delta \iint L(u_t, u_x, u) \, dt \, dx = 0,$$

where

$$(6.2) \qquad L = \tfrac{1}{2} u_t^2 - \tfrac{1}{2} u_x^2 - V(u).$$

The periodic solution takes the form $u = U(\theta)$, where $\theta_x = k$ and $\theta_t = -\omega$ are constants, and the actual solution for $U(\theta)$ will bring in another constant A, which is equivalent to the amplitude. We now consider trial functions of this form. The Lagrangian becomes

$$(6.3) \qquad L^{(0)} = L(-\omega U_\theta, kU_\theta, U).$$

Variations of this with respect to U must yield the equation for the periodic solution. For the Klein-Gordon example,

$$(6.4) \qquad L^{(0)} = \tfrac{1}{2}(\omega^2 - k^2)U_\theta^2 - V(U),$$

and, indeed, (2.24) follows from its variational equation

$$(6.5) \qquad \frac{d}{d\theta}\left(\frac{\partial L^{(0)}}{\partial U_\theta}\right) - \frac{\partial L^{(0)}}{\partial U} = 0.$$

But our aim is to go further and find equations for (ω, k, A) when they are allowed to vary. We now describe a procedure based on intuitive arguments; its formal justification will be provided later. For the uniform wave train, calculate the "average" Lagrangian

$$(6.6) \qquad \mathcal{L}(\omega, k, A) = \frac{1}{2\pi} \int_0^{2\pi} L(-\omega U_\theta, kU_\theta, U) \, d\theta,$$

in terms of the constant parameters (ω, k, A). (There is some subtlety as to how this should be done, which we return to shortly.) We propose that in the extension to slowly varying wave trains, the slowly varying functions ω, k, A will satisfy the variational equations given by the "averaged" variational principle

$$(6.7) \qquad \delta \iint \mathcal{L}(\omega, k, A) \, dx \, dt = 0.$$

In this variational principle, ω and k cannot be varied independently; they are related by

$$\omega = -\theta_t, \qquad k = \theta_x.$$

The variational equations resulting from variations of δA and $\delta \theta$ in (6.7) are

$$(6.8) \quad \delta A: \qquad\qquad\qquad \mathcal{L}_A = 0,$$

$$(6.9) \quad \delta\theta: \qquad\qquad \frac{\partial}{\partial t} \mathcal{L}_\omega - \frac{\partial}{\partial x} \mathcal{L}_k = 0,$$

respectively. Once (6.9) has been obtained, it is convenient to work with ω and k, and complete the system with the consistency relation

$$(6.10) \qquad\qquad \frac{\partial k}{\partial t} + \frac{\partial \omega}{\partial x} = 0.$$

This is the whole theory. Equation (6.8) is an equation relating ω, k, A, so it *must* be the dispersion relation. Equation (6.9) must be the sought-after differential equation for the amplitude. We now consider the content in detail.

LINEAR PROBLEMS. For any linear problem, the Lagrangian must be quadratic in u and its derivatives. The periodic solution may be taken in the form

$$(6.11) \qquad\qquad U = a \cos \theta.$$

Hence, when this is substituted in (6.6), the average Lagrangian *must* be proportional to a^2 and take the form

(6.12) $$\mathcal{L} = G(\omega, k)a^2.$$

Then (6.8) becomes

(6.13) $$G(\omega, k) = 0.$$

Therefore, without detailed calculation, we know that the function appearing in (6.12) *must* be the linear dispersion function. It is interesting to note that the stationary value is $\mathcal{L} = 0$. For those cases in which $\mathcal{L}$ is the difference of kinetic and potential energy, this shows the equipartition of energy between the two. It is also clear that the dispersion relation itself should not be used in substituting the periodic trial functions (6.11). This would lead to $\mathcal{L} = 0$; we would already have the stationary value, and there would not be enough freedom to apply the variational argument.

The amplitude equation (6.9) becomes

(6.14) $$\frac{\partial}{\partial t}(G_\omega a^2) - \frac{\partial}{\partial x}(G_k a^2) = 0.$$

If a solution of (6.13) is $\omega = W(k)$, then $G(W, k) = 0$, and

$$G_\omega W'(k) + G_k = 0.$$

Hence, the group velocity C can be written in terms of G as

(6.15) $$C = -\frac{G_k}{G_\omega}.$$

Thus, (6.14) takes the form

(6.16) $$\frac{\partial}{\partial t}\{g(k)a^2\} + \frac{\partial}{\partial x}\{g(k)C(k)a^2\} = 0.$$

In this equation, the factor $g(k)$ can be slipped out, for the equation may be expanded to

(6.17) $$g(k)\left\{\frac{\partial(a^2)}{\partial t} + \frac{\partial}{\partial x}(Ca^2)\right\} + g'(k)\left\{\frac{\partial k}{\partial t} + C(k)\frac{\partial k}{\partial x}\right\} = 0.$$

The last term vanishes from (6.9), so we have

$$(6.18) \qquad \frac{\partial (a^2)}{\partial t} + \frac{\partial}{\partial x} \left\{ C(k) a^2 \right\} = 0.$$

This is the general proof of (4.7). The identification of (6.9) and its relation with the energy equation is left until the nonlinear case has been discussed.

KLEIN-GORDON EXAMPLE. In nonlinear problems, the manipulations to obtain the function $\mathcal{L}$ are more subtle, and it is as well to do a specific example before the general case. For the nonlinear Klein-Gordon equation, the averaged Lagrangian is first obtained from (6.4) and (6.6) in the form

$$(6.19) \qquad \mathcal{L} = \frac{1}{2\pi} \int_0^{2\pi} \left\{ \frac{1}{2} \left(\omega^2 - k^2 \right) U_\theta^2 - V(U) \right\} d\theta.$$

The periodic solution satisfies (2.25). This is used to write (6.19), successively, as

$$(6.20) \qquad \mathcal{L} = \frac{1}{2\pi} \int_0^{2\pi} (\omega^2 - k^2) U_\theta^2 \, d\theta - A$$

$$= \frac{1}{2\pi} \int_0^{2\pi} (\omega^2 - k^2) U_\theta \, dU - A$$

$$= \frac{1}{2\pi} \sqrt{2(\omega^2 - k^2)} \oint \sqrt{A - V(U)} \, dU - A,$$

where again, $\oint$ denotes the integral over a complete period of the integrand. The final form does not require the dependence of U on θ.

According to (6.8), the dispersion relation should be

$$\mathcal{L}_A = \frac{1}{2\pi} \sqrt{\frac{1}{2}(\omega^2 - k^2)} \oint \frac{dU}{\sqrt{A - V(U)}} - 1 = 0,$$

and, correctly, this checks with (2.27). In the linear limit, $V = \frac{1}{2} U^2$, and the dependence on A drops out. In fact, in this limit, (6.20) becomes

$$\mathcal{L} = (\sqrt{\omega^2 - k^2} - 1) A;$$

since A is then proportional to a^2, this conforms to the general arguments.

HAMILTONIAN TRANSFORMATION. The transformation from (6.19) to (6.20) is not only a matter of convenience. The introduction of A into (6.19) uses information from the first integral (2.25) of the periodic solution, and yet the form of the trial functions $U(\theta)$ must be left flexible enough to allow the variational argument to go through. For example, (2.25) implicitly contains the dispersion relation (as was seen in (2.27)), and we do not want to introduce that explicitly; it should be a consequence of the variational principle. This is an important point that can only be fully explained by following the formal expansion procedure described in Sec. 9. Here, it is essential that the procedure should be unambiguous at least. It is the following. The periodic solution is given by (6.5), which always has the first integral

$$(6.21) \qquad U_\theta \frac{\partial L^{(0)}}{\partial U_\theta} - L^{(0)} = A.$$

Define a new variable

$$(6.22) \qquad \Pi = \frac{\partial L^{(0)}}{\partial U_\theta},$$

in place of U_θ, and let

$$(6.23) \qquad H(\Pi, U) = U_\theta \frac{\partial L^{(0)}}{\partial U_\theta} - L^{(0)}.$$

This is like the Hamiltonian transformation in mechanics. The integral (6.21) is then

$$(6.24) \qquad H(\Pi, U) = A.$$

Now, from (6.21),

$$(6.25) \qquad \mathcal{L} = \frac{1}{2\pi} \int_0^{2\pi} L^{(0)} \, d\theta$$

$$= \frac{1}{2\pi} \int_0^{2\pi} \{\Pi U_\theta - A\} \, d\theta$$

$$= \frac{1}{2\pi} \oint \Pi \, dU - A.$$

Finally, Π can be determined from (6.24) as a function of U and the parameters ω, k, A to write the loop integral as a function of ω, k, and A.

HIGHER-ORDER SYSTEMS, NONUNIFORM MEDIA.
The theory goes through in a similar way when the Lagrangian in (6.1) contains more than one variable u and higher derivatives than the first. There is one feature that is added with more dependent variables. It turns out that some of the additional variables appear only through their derivatives; they are potentials whose derivatives are the significant physical quantities. If φ denotes such a variable, the most general periodic solution for the physical quantities will require φ to be taken in the form

$$\varphi = \beta x - \gamma t + \Phi(\theta),$$

where $\Phi(\theta)$ is periodic. The β and γ are extra parameters in the solution. In the extension to slowly varying wave trains, the quantity $\beta x - \gamma t$ has to be treated in the same way as $\theta = \kappa x - \omega t$; i.e., we must take

$$\varphi = \psi(x, t) + \Phi(\theta)$$

and define β, γ by

$$\beta = \psi_x, \qquad \gamma = -\psi_t.$$

These derivatives of the function ψ will appear in the averaged Lagrangian, and variations $\delta\psi$ will give the equation

$$\frac{\partial}{\partial t} \mathcal{L}_\gamma - \frac{\partial}{\partial x} \mathcal{L}_\beta = 0$$

in analogy to (6.9). Other constants of integration similar to A lead to equations like (6.8) and provide the appropriate additional relations between them.

The extensions to more space variables are immediate. The wave number becomes a vector $k_i = \theta_{x_i}$, and the variation of (6.7) with respect to θ gives

$$\frac{\partial}{\partial t} \mathcal{L}_\omega - \frac{\partial}{\partial x_i} \mathcal{L}_{k_i} = 0,$$

as the extension of (6.9).

For nonuniform media, the original Lagrangian in (6.1) will have also explicit dependence on $\mathbf{x}$ or t, through the parameters that describe the medium. The periodic solution and the average Lagrangian $\mathcal{L}$ are calculated by holding those parameters fixed. They are then freed in (6.7); that is,

$$\mathcal{L} = \mathcal{L}(\omega, \mathbf{k}, A; \mathbf{x}, t).$$

The extra dependence on $\mathbf{x}$ or t does not affect the variational argument, and equations (6.8) and (6.9) apply. Since the dispersion relation involves $\mathbf{x}$ or t, the coefficients in (6.14) will do also, and equations such as (6.18) will pick up additional terms.

The single inclusive form to cover all these various possibilities is one of the successes of the theory.

7. ADIABATIC INVARIANTS, WAVE ACTION, ENERGY

It is interesting, both for its own sake and as a guide to identifying equation (6.9), to note that the present approach can be used in the corresponding problems of oscillating systems in dynamics and provides a different derivation of some of the standard results there. In dynamics, the governing equations are ordinary differential equations for functions of t, and the present theory can be applied by dropping the dependence on x. Periodic solutions can then be modulated only through a slowly varying parameter, $\lambda(t)$ say, and the Lagrangian in (6.1) takes the form $L(q_t, q, \lambda(t))$. A nonlinear oscillator with $L = \frac{1}{2}q_t^2 - V(q, \lambda)$ would be an example closely related to (6.2).

The analysis proceeds exactly as described in Sec. 6, with the x dependence omitted at each step. The results are

(7.1) $$\frac{d}{dt}\mathcal{L}_\omega = 0, \qquad \mathcal{L}_E = 0,$$

and

(7.2) $$\mathcal{L} = \frac{\omega}{2\pi} \oint p \, dq - E,$$

where p is the usual generalized momentum and $p(\lambda, E)$ is determined from the energy integral

(7.3) $$H(p, q, \lambda) = E.$$

(In this case the quantity Π introduced in (6.22) reduces to $\Pi = \omega p$, and it is more convenient to use the usual momentum p.) The first result in (7.1) shows that

$$(7.4) \qquad I(\lambda, E) = \mathcal{L}_\omega = \frac{1}{2\pi} \oint p \, dq$$

is constant for the slow variations; it is the so-called "adiabatic invariant." Its expression as the loop integral $\oint p \, dq$ is well-known, but the expression as $\mathcal{L}_\omega$ is new. Of course, the energy E is not constant as the parameter λ varies. The second relation in (7.1) determines the frequency ω by

$$\omega \frac{\partial I}{\partial E} = 0,$$

which is also well-known.

In the waves problem, the adiabatic equation becomes the conservation equation (6.9)

$$\frac{\partial}{\partial t} \mathcal{L}_\omega - \frac{\partial}{\partial x} \mathcal{L}_k = 0.$$

There is a nice duality between the quantities $\mathcal{L}_\omega$ and $\mathcal{L}_k$. If the wave train is uniform in x, but the medium changes in t, $\mathcal{L}_\omega =$ constant; if a periodic wave train enters a region where the medium varies with x, $\mathcal{L}_k =$ constant. In the propagation of modulations along a wave train, changes of $\mathcal{L}_\omega$ in time are balanced by changes of $\mathcal{L}_k$ in space. This is the conservation of something, and it is convenient to call it "wave action"; $\mathcal{L}_\omega$ and $-\mathcal{L}_k$ are the density and flux of wave action.

The application in dynamics has shown the crucial distinction between this equation and the energy equation. This carries over to the case of waves. The energy equation is obtained from considerations of time-invariance in the variational principle. It is

$$(7.5) \qquad \frac{\partial}{\partial t} (\omega \mathcal{L}_\omega - \mathcal{L}) - \frac{\partial}{\partial x} (\omega \mathcal{L}_k) = -\mathcal{L}_t.$$

This may be established directly from (6.8), (6.9), and (6.10), but the derivation from time-invariance identifies it as the energy equation. For a medium independent of time, the right-hand side is zero, and this equation is also in conservation form.

For linear problems, since $\mathcal{L}$ is quadratic in a, this equation also takes the form (6.16), but with a different function for g. For a uniform medium, the g is a function of k and may be eliminated as before to give (6.18). Thus, (6.18) *is* related to the energy equation, but the wave action is both more fundamental and more convenient for nonhomogeneous media.

We may note one further point in linear problems; the stationary value is $\mathcal{L} = 0$, so the energy density

$$E = \omega \mathcal{L}_\omega.$$

Hence, the wave action equation may be written

$$(7.6) \qquad \frac{\partial}{\partial t}\left(\frac{E}{\omega}\right) + \frac{\partial}{\partial x}\left(\frac{CE}{\omega}\right) = 0.$$

The expression of the adiabatic invariant as E/ω for linear problems in mechanics is, of course, well-known. Reference [12] claims some difficulty in applying these ideas to moving media, due to mysteries in the transformation between moving frames. There is none. In the reference frame moving with the medium, an observer would use

$$\mathcal{L}_0 = G_0(\omega_0, k)a^2,$$

say, whereas an observer who takes the medium to have a velocity V will use

$$\mathcal{L} = G(\omega, k)a^2,$$

where the relation between them is

$$\omega = \omega_0 + Vk, \qquad G(\omega, k) = G_0(\omega - Vk, k).$$

In the first frame

$$E_0 = \omega_0 \mathcal{L}_{\omega_0} = \omega_0 G_{0\omega_0}a^2,$$

and in the second frame

$$E = \omega \mathcal{L}_\omega = \omega G_\omega a^2.$$

Since

$$G(\omega, k) = G_0(\omega - Vk, k),$$

it follows that

$$C = -\frac{G_k}{G_\omega} = V - \frac{G_{0_k}}{G_{0_\omega}} = V + C_0$$

and

$$\frac{E}{\omega} = \frac{E_0}{\omega_0}.$$

Thus, (7.6) is correct but could also, more usefully, be rewritten

$$\frac{\partial}{\partial t}\left(\frac{E_0}{\omega_0}\right) + \frac{\partial}{\partial x}\left\{(V + C_0)\frac{E_0}{\omega_0}\right\} = 0.$$

The momentum equation may also be derived from the variational principle; it is derived from considerations of space invariance. It is

$$-\frac{\partial}{\partial t}(k\mathcal{L}_\omega) + \frac{\partial}{\partial x}(k\mathcal{L}_k - \mathcal{L}) = -\mathcal{L}_x,$$

which may also be verified directly from (6.8), (6.9), and (6.10).

8. NONLINEAR GROUP VELOCITY; STABILITY OF PERIODIC WAVES

For linear problems, the group velocity is the characteristic velocity for (6.10) and appears in the calculation of the distribution of $k(x, t)$. With k known, it appears a second time as the characteristic velocity in solving (6.14) [or (6.18)] for $a(x, t)$. For nonlinear problems the equations (6.9) and (6.10) for k and a are coupled together. If these form a hyperbolic pair, it is natural to extend the definition of group velocity to the characteristic velocities of this pair. The two velocities are no longer equal for nonlinear problems. This will make important differences in the propagation of a modulation down the wave train. For a nonlinear wave train, an initially concentrated disturbance should eventually split into two disturbances. It would be of great interest to see whether this can be observed.

If the equations (6.9) and (6.10) are elliptic, it is easy to deduce that perturbations in x grow exponentially in time. This shows that the periodic wave train is itself unstable. Coincidentally with the development of this theory, Dr. T. B. Benjamin [13] had deduced that *deep* water waves were unstable in this sense. He noted that experiments to produce a periodic wave train always showed instability after several wavelengths from the wave maker, and did an ingenious "side-band" instability analysis, working with the second approximation beyond linear theory to explain it. The present analysis was then used for waves of moderate amplitude in water of *finite* depth h, [10]. It showed that the equations were elliptic for $kh > 1.36$ and hyperbolic for $kh < 1.36$. Thus water waves of finite amplitude should be unstable for $kh > 1.36$.

The analysis for moderate amplitude is very simple when the only parameters are (ω, k, a); it becomes more involved when further parameters such as the β and γ referred to in Sec. 6 arise. Deep water waves do involve only (ω, k, a), but the finite depth case requires additional parameters, which refer to small but important changes in mean depth and mean mass flow produced by the waves. The simpler case is presented here. In such problems, with moderately small amplitude, the main coupling between (6.9) and (6.10) is through the dependence of ω on a in the dispersion relation. If this is written to first order in a^2 as

$$(8.1) \qquad \omega = \omega_0(k) + \omega_1(k)a^2,$$

it turns out to be sufficient to use

$$(8.2) \qquad \frac{\partial k}{\partial t} + \frac{\partial}{\partial x}\{\omega_0(k) + \omega_1(k)a^2\} = 0,$$

and retain (6.18) in the linear approximation

$$(8.3) \qquad \frac{\partial a^2}{\partial t} + \frac{\partial}{\partial x}\{C_0(k)a^2\} = 0, \qquad C_0 = \omega_0'(k).$$

A simple calculation shows that the characteristic velocities are

$$(8.4) \qquad C = C_0 \pm a\sqrt{\omega_1 C_0'} + O(a^2).$$

If $\omega_1 C_0' > 0$, the double characteristic velocity C_0 splits into the two real roots given by (8.4), the system is hyperbolic, and (8.4) gives the two nonlinear group velocities. If $\omega_1 C_0' < 0$, the system is elliptic, and the wave train is unstable. For deep water waves,

$$\omega = \sqrt{gk}\,(1 + \tfrac{1}{2}k^2a^2) + O(a^4);$$

hence $\omega_1 = \frac{1}{2}g^{1/2}k^{5/2} > 0$, $C_0' = \omega_0'' = -\frac{1}{4}g^{1/2}k^{-3/2} < 0$, and the wave train is unstable. Full details and the relations with Benjamin's analysis are given in [2] and [10].

9. FORMAL PERTURBATION THEORY

In this final section a brief account is given of how the theory presented in Sec. 6 may be formally justified as the first term in a perturbation procedure. A full account is given in [11].

The method turns out to be an interesting extension of the so-called "two-timing method," which is usually applied directly to differential equations, but is here adapted to the variational principle. The method recognizes explicitly in the dependent variables that changes are occurring on two time scales: the fast oscillations of the wave train and the slow variations of the parameters (ω, κ, a). There are two corresponding length scales. The function $u(x, t)$ is expressed in the form

$$u(x, t) = U(\theta, X, T),$$

where

$$\theta = \epsilon^{-1}\Theta(X, T), \qquad X = \epsilon x, \qquad T = \epsilon t,$$

and the small parameter ϵ measures the ratio of the fast time scale to the slow time scale; the function U no longer refers just to the periodic solution. If the wave number k and frequency ω are introduced, we have

$$(9.1) \qquad \omega(X, T) = -\theta_t = -\Theta_T, \qquad k(X, T) = \Theta_X.$$

The scaling has been arranged so that

$$\frac{\partial \omega}{\partial t} = \epsilon \frac{\partial \omega}{\partial T}, \qquad \frac{\partial \omega}{\partial x} = \epsilon \frac{\partial \omega}{\partial X},$$

with similar expressions for k, and

$$\frac{\partial u}{\partial t} = -\omega \frac{\partial U}{\partial \theta} + \epsilon \frac{\partial U}{\partial T}, \qquad \frac{\partial u}{\partial x} = k \frac{\partial U}{\partial \theta} + \epsilon \frac{\partial U}{\partial X}.$$

If X, T, ω, k, U are all taken to be $O(1)$ quantities, the scaling has been arranged so that ω, k are slowly varying, and so that u has a slow variation in addition to its oscillation with the phase θ.

The Euler equation for (6.1) is

$$\frac{\partial}{\partial t} L_1 + \frac{\partial}{\partial x} L_2 - L_3 = 0,$$

where

$$L_1 = \frac{\partial L}{\partial u_t}, \qquad L_2 = \frac{\partial L}{\partial u_x}, \qquad L_3 = \frac{\partial L}{\partial u}.$$

In the scaled variables, this becomes

(9.2) $$-\omega \frac{\partial}{\partial \theta} L_1 + \kappa \frac{\partial}{\partial \theta} L_2 - L_3 + \epsilon \frac{\partial L_1}{\partial T} + \epsilon \frac{\partial L_2}{\partial X} = 0,$$

where the arguments in the L_j are

(9.3) $$L_j = L_j(-\omega U_\theta + \epsilon U_T, kU_\theta + \epsilon U_X, U).$$

Equation (9.2) is written as a second-order equation for the *three-variable* function $U(\theta, X, T)$. The art of two-timing technique is to solve this equation by treating θ, X, T as *independent* variables. If this can be done, then clearly $U(\epsilon^{-1}\theta, X, T)$ is a solution of the original problem. The solution is usually obtained by means of an expansion

$$U(\theta, X, T) = \sum_{n=0}^{\infty} \epsilon^n U_n(\theta, X, T),$$

and the extra flexibility is used to suppress secular terms in θ, which would otherwise limit the uniform validity of the expansion. To lowest order in ϵ we have

$$\frac{\partial}{\partial \theta} (-\omega L_1^{(0)} + kL_2^{(0)}) - L_3^{(0)} = 0,$$

where

$$L_j^{(0)} = L_j(-\omega U_{0\theta}, kU_{0\theta}, U_0).$$

It has the first integral

(9.4) $$(-\omega L_1^{(0)} + kL_2^{(0)})U_{0\theta} - L^{(0)} = A(X, T).$$

Comparing with (6.21), we see that this is just the equation for periodic solution but with the added dependence on (X, T). At this lowest order, therefore, we find that U_0 has the same form as the periodic solution, but that the parameters ω, k, A are automatically functions of (X, T) to describe the slow variations.

The equations for (ω, k, A) may then be found by proceeding to the next order in ϵ. An equation for U_1 is found, and the required equations for (ω, k, A) may be derived from the conditions to suppress a secular term proportional to θ in U_1.

However, our interest is in the variational principle. Now it is a surprising fact that (9.2) is the Euler equation for the variational principle

$$(9.5) \quad \delta \iiint_0^{2\pi} L(-\omega U_\theta + \epsilon U_T, \, kU_\theta + \epsilon U_X, \, U) \, dT \, dX \, d\theta = 0,$$

for the three-variable function $U(\theta, X, T)$. In (9.5) the function U and its variations are taken to be periodic in θ, and the variations in U vanish on the boundary of the (X, T) region. If we define

$$\overline{L} = \frac{1}{2\pi} \int_0^{2\pi} L(-\omega U_\theta + \epsilon U_T, \, kU_\theta + \epsilon U_X, \, U) \, d\theta,$$

we have already an exact form of the "averaged" variational principle. Not only is the intuitive idea sound as a first approximation, it contains the whole expansion. To lowest order,

$$\overline{L} = \mathfrak{L} = \frac{1}{2\pi} \int_0^{2\pi} L(-\omega U_{0\theta}, \, kU_{0\theta}, \, U_0) \, d\theta,$$

where U_0 is the periodic solution extended to allow ω, k, A to depend on X, T as in (9.4). The quantity $\mathfrak{L}$ is calculated as a function of (ω, k, A) alone, as described in Sec. 6. We then have that, to lowest order, (9.5) is

$$(9.6) \qquad \delta \iint \mathfrak{L}(\omega, k, A) \, dX \, dT = 0,$$

where

$$\omega = -\Theta_T, \qquad k = \Theta_X.$$

The variations in Θ and A give

$$\frac{\partial}{\partial T} \mathfrak{L}_\omega - \frac{\partial}{\partial X} \mathfrak{L}_k = 0,$$

$$\mathfrak{L}_A = 0,$$

respectively.

This shows the main ideas in the perturbation scheme. There are a number of points that would require longer explanation than is appropriate here. They are covered in the complete account in [11].

This research was supported by the Office of Naval Research, U.S. Navy.

REFERENCES

1. Lamb, H., *Hydrodynamics*, 6th ed. Cambridge: Cambridge University Press, 1932, Chap. 9.

2. Whitham, G. B., *Proc. Roy. Soc.*, A **299** (1967), 6.

3. Seliger, R. L., *Proc. Roy. Soc.*, A **303** (1968), 493.

4. Jeffreys, H., and B. S. Jeffreys, *Methods of Mathematical Physics*, 3rd ed. Cambridge: Cambridge University Press, 1956, Chap. 17.

5. Snodgrass, F. E., et al., *Phil. Trans. Roy. Soc.*, A **259** (1966), 431.

6. Whitham, G. B., *Communications Pure and Applied Math.*, **14** (1961), 675.

7. Mowbray, D. E., and B. H. S. Rarity, *J. Fluid Mechanics*, **28** (1967), 1.

8. Whitham, G. B., *J. Fluid Mechanics*, **22** (1965), 273.

9. ———, *Proc. Roy. Soc.*, A **283** (1965), 238.

10. ———, *J. Fluid Mechanics*, **27** (1967), 399.

11. ———, *J. Fluid Mechanics*, 44 (1970), 373.

12. Bretherton, F. P., and C. J. R. Garrett, *Proc. Roy. Soc.*, A **302** (1969), 529.

13. Benjamin, T. B., *Proc. Roy. Soc.* A **299** (1967), 59.

INDEX